AF556952

Controlling Multivariable Processes

An Independent Learning Module from the Instrument Society of America

CONTROLLING MULTIVARIABLE PROCESSES

By F. G. Shinskey

THE FOXBORO COMPANY
Foxboro, Massachusetts, U.S.A.

INSTRUMENT SOCIETY OF AMERICA

Printed in the United States of America

INSTRUMENT SOCIETY OF AMERICA
67 Alexander Drive
P.O. Box 12277
Research Triangle Park
North Carolina 27709

LIBRARY OF CONGRESS CATALOGING IN PUBLICATION DATA

LIBRARY OF CONGRESS
CATALOG CARD NO.:81-81497

Shinskey, F. G.
Controlling Multivariable Processes

Research Triangle Park, N.C.: Instrument
Society of America

214 p.

8111 810403

ISBN: 0-87664-529-5

Editorial development and book design by Monarch International, Inc.
Under the editorial direction of Paul W. Murrill, Ph.D.

Production by Publishers Creative Services, Inc.

Second Printing April 1983

Table of Contents

Preface

ISA's Independent Learning Modules

This is an Independent Learning Module (ILM) on *Controlling Multivariable Processes;* it is part of the ISA Series of Modules on Control Principles and Techniques.

Comments about This Volume

This ILM on *Controlling Multivariable Processes* is intended to guide the reader in the procedures which are essential in applying controls to modern industrial processes. Many of the procedures are mathematical in nature—unavoidably so, in that the resulting systems are implementations of mathematical solutions. Yet most of the procedures involve simple algebra, with partial differentiation being the most demanding discipline.

The course is intended for engineers who are responsible or will be responsible for control-system application and design. A fundamental understanding of feedback control is an essential prerequisite; the ILM on *Fundamentals of Process Control Theory* will be helpful in this regard. Engineers having a process background will find this module easier to follow than those without because system designs are based on mathematical relationships that describe the process to be controlled. Nonetheless the presentation, examples, and exercises will be helpful to all who wish to develop skills in control-system design.

Unit 1: Introduction

UNIT 1

Introduction

Welcome to ISA's Independent Learning Module *Controlling Multivariable Processes*. This first unit presents the general control problem for the process plant, its multivariable nature, and the need for coordinating these variables.

Learning Objectives — When you have completed this unit, you should:

A. Understand the multivariable nature of production processes.

B. Be able to classify the variables into broad groups.

C. Appreciate the need for coordinating their controls into a system structure.

1-1. Process Inputs and Outputs

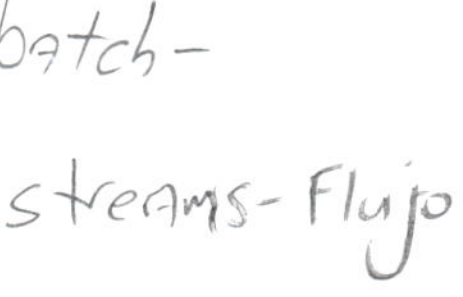

Any process may be represented as a block, with variables both entering and leaving. The input variables are acted upon by the process to develop the output variables as in Fig. 1-1. Input variables are typically rates of flow in a continuous process, or quantities in a batch process. These streams may be either entering or leaving the process—stream direction does not determine the flow of information. (Observe that flow out of a vessel affects its inventory to the same degree as inflow.) Some input variables are capable of independent manipulation by an operator or by a controller—these are called *manipulated* variables. The remaining input variables, which cannot be independently manipulated, are considered to be *disturbing* variables. Often these are the output variables from other processes, in which case they may be either controlled or uncontrolled. Disturbing variables may also include the conditions surrounding the process, such as ambient temperature.

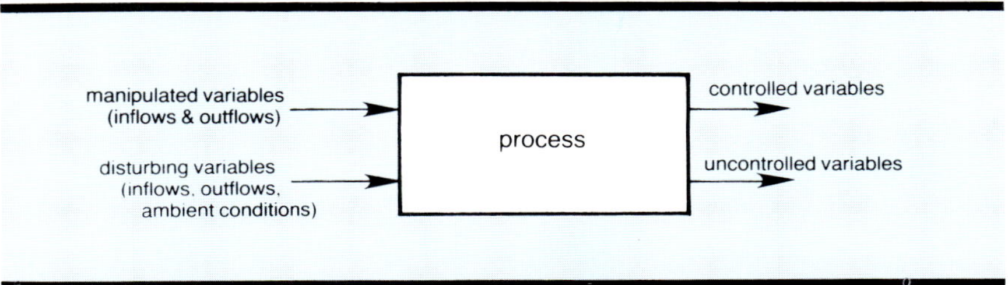

Fig. 1-1. Process inputs and outputs are designated on the basis of information flow.

Process output variables are considered either *controlled* or *uncontrolled*. Control is achieved through manipulation, so that each variable that is controlled requires the manipulation of a process input. Output variables that are uncontrolled simply may not be influenced by any of the manipulable inputs. Also, the number of output variables exceeding the number of manipulated input variables must remain uncontrolled.

The combination of a single controlled variable and a single manipulated variable is considered a single-loop process, regardless of the number of disturbing and uncontrolled variables present. There are many familiar single-loop processes—controlling room temperature with a steam valve and controlling the pH of a waste stream by adding reagent are two of the more common examples.

1-2. Multiple Controlled Variables

Even the simplest of production processes requires the control of *two* variables—product rate and quality. An example of a process with two controlled variables would be the blending of two streams to form a mixture as in Fig. 1-2. Product flow must satisfy certain market demands in the long term and productivity goals in the short term. Secondly, the composition of the mixture must meet some specification in order to satisfy the use for which it is intended. If multiple specifications must be met, additional product-quality control loops must be added to the basic two-variable system.

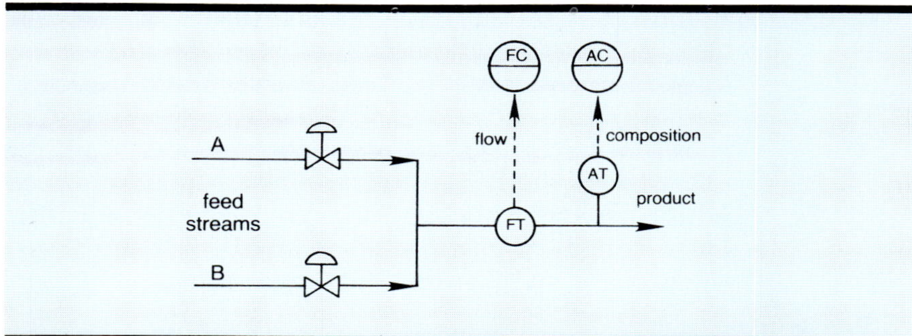

Fig. 1-2. The two-stream blender is perhaps the simplest multivariable process.

Usually a feedstock must pass through a series of stage-wise operations in its conversion into a finished product. Control may be required at each stage to obtain certain desired characteristics in the product. Most processes have storage vessels located between these sequenced operations to give a degree of dynamic isolation between them. This practice adds loops for inventory control, even when the storage volumes are minimal.

Separation processes split a feed stream into two or more products, each of which may require quality control. These processes are driven by a flow of energy whose balance must be maintained. This is the type of operation common to oil refineries, where a crude feedstock containing many components is separated into many products with fewer components.

Modern processing plants have both series and parallel operations interconnected by flow networks that can become quite complex. As a consequence, virtually every process encountered by the control engineer is multivariable in nature. While his knowledge of single-loop control is essential in applying controls to these processes, it is by no means sufficient.

1-3. Types of Controlled Variables

The types of processes encountered in the industrial world are too many to be meaningfully classified. But, as a beginning, the controlled variables themselves may be readily classified, giving an insight into their role and importance in the process, and identifying the means commonly used to measure them. Table 1-1 lists five categories of controlled variables, along with the measurements that usually identify them. Because most of the measurements appear more than once in the table, an understanding of their roles in the functioning of the process is vital to success of the control system. For example, an environmental temperature-control loop contributes a different function to the process than an inventory temperature-control loop. Hence, it may operate differently and be subject to different forces. Each category is examined in detail in a later unit, but at this point they are introduced for purposes of distinction.

Variable type	Measurement
1. Production rate	Flow
2. Inventory	
a. Gas	Pressure
b. Liquid	Liquid level or pressure
c. Solid	Weight
d. Composition	Chemical analysis
e. Energy	Temperature or pressure
3. Environmental	
a. Temperature	Temperature
b. Pressure	Pressure
c. Composition	Chemical analysis
4. Product quality	
a. Physical	Physical property
b. Chemical	Chemical analysis
5. Economic	Flow, flow ratio, valve position, chemical analysis

Table 1-1. Classification of Controlled Variables

The production-rate variable determines the rate at which a process operates. There is generally only one of these in any given process. It is usually the rate of feed, but in some cases it may be the rate of flow of a final or intermediate product. When operations are conducted in series, a single production rate sets the steady-state throughput of all operations. Production rate is a disturbing variable to all processes because it affects all other controlled variables.

Inventory variables represent the accumulation of material or energy at specific points in the process. In the steady state, inventory must remain constant (this defines the steady state); hence, inventory controllers are responsible for closing material and energy balances. But a steady state can be maintained at *any* level of inventory. Therefore, one issue that will be discussed in Unit 3 is the choice of an appropriate inventory level.

Environmental variables determine the conditions under which a process functions. They are particularly important in conducting chemical reactions, which are sensitive to all three dimensions —temperature, pressure, and composition. Room temperature is an environmental variable affecting the process of people at work. Environmental variables tend to affect process performance as much as product quality.

Product quality needs to be controlled in every process. It is perhaps the most important controlled variable because product value depends on it. It is the most difficult to control because it is disturbed by almost every other variable in the process. It also is difficult to measure in most cases, usually requiring sampling and even off-line determinations, whose delays degrade control.

The last group of controlled variables is economic in nature, and in fact, these are rarely controlled. They represent the cost of production, which is usually not obtainable in a single measurement. Occasionally, losses are measurable as the flow of a costly agent or premium fuel needed for control, or as the concentration of valuable product in a waste stream. Economic controls are being added to conserve energy and other resources, but only after product-quality goals have been satisfied.

Example 1-1

Classify the controlled variables represented by the transmitted signals from the process in Fig. 1-3, into the five categories given in Table 1-1.

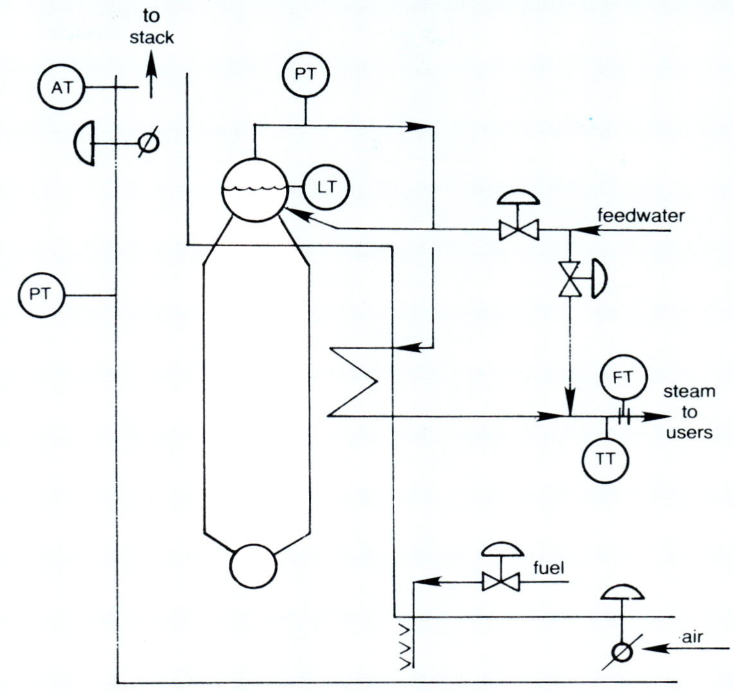

Fig. 1-3. A superheated steam boiler features each of the five classes of controlled variables.

1. Production rate is steam flow, usually determined by demands placed on the steam header by users. In this illustration it is an uncontrolled variable or controlled elsewhere.

2. Inventory of liquid is drum level—as much feedwater must be supplied as steam is boiled away to keep liquid inventory constant.

   Inventory of energy is steam pressure—heat removed in the steam must be supplied by the combustion of fuel.

3. The environmental variable is furnace pressure or draft, which is controlled to maintain efficient and safe firing.

4. Product quality is the temperature of the superheated steam.

5. The economic variable in this example is the composition of the flue gas, as this determines how much energy is lost to the stack.

1-4. Process Interactions

The principal challenge in designing control systems for industrial processes is arriving at a structure which minimizes the interactions between variables. If interactions were absent, control would be unnecessary. As it is, interactions always are present, and to varying degrees.

Production rate influences all other variables, because production requires the coordination of many resources into the product. As production rate increases, the rates of all resources entering the process, i.e. feedstocks, solvents, fuels, work, etc., must increase proportionately if product quality is to be maintained.

If production rate is changed unilaterally, quality and inventory variables tend to be upset first. The actions of their controllers in responding to the disturbance in turn disturb other variables.

Consider the boiler of Fig. 1-3. Should the user of steam increase its withdrawal rate, steam pressure and temperature will begin to fall almost immediately due to a reduction in energy inventory. Liquid level will follow. To return pressure to its control point, fuel flow must increase. This has two secondary effects—liquid level will fall faster as the rate of boiling increases, and the flue gas will become rich in unburned fuel if air flow is not increased. Adjusting air flow can correct the oxygen-deficient condition, but this action upsets furnace pressure.

The existence of process interactions poses two problems to the control-systems engineer:

1. With most processes it is impossible to connect pairs of manipulated and controlled variables into single loops that will not interact. However, there is usually a "best" pairing for any given process, and determining this pairing is essential to achieving steady-state stability in the process.

2. Interaction tends to be so pervasive that even the "best" configuration of single loops is a compromise that may fail to give acceptable dynamic performance. In most cases then, substantial improvement can be gained by coordinating the variables in relationship to their natural interaction within the process.

The magnitude of the system-design problem varies factorially with the number of manipulated variables. (The case in which there is an equal number of controlled and manipulated variables is clear, because that is also the number of possible independent single loops.) Where there are only two manipulated and two controlled variables, as in Fig. 1-2, there are only two possible configurations of single loops: Valve A can be used to control variable 1 and valve B variable 2, or valve A can control variable 2 and valve B variable 1. The engineer can usually make the best assignment intuitively. For the system in Fig. 1-2, he would rightly manipulate the larger stream to control flow, and the smaller to control composition.

However, the choice is not alway obvious, and occasionally the wrong pairing is chosen. But in this 2×2 case, the problem is easily corrected by exchanging the connections—there are only two possible configurations, and if one is unsatisfactory, the other must be better.

The problem quickly explodes in size as more variables are added. A typical distillation column has five manipulated variables, offering the possibility of 120 different configurations of five single loops. Many of these configurations will have one or more loops that are intuitively ineffective and therefore can be eliminated from consideration. Yet many possibilities remain, and the optimum selection is far from obvious. In fact, it is only through careful analysis of the process interactions that the optimum configuration can be found. Even then, certain equipment constraints may preclude implementation of the optimum, and a suboptimum configuration may have to be accepted.

An examination of the boiler in Fig. 1-3 illustrates the ineffectiveness of single-loop controls in coping with interaction. Flue-gas oxygen content, for example, is affected equally by fuel and air. Therefore it is not only foolish but dangerous to manipulate them independently. The most effective control systems will then not be a simple arrangement of single loops, but will be a coordinated structure, whose interactions mirror the relationships which naturally exist in the controlled process. Increases in process performance and efficiency are bought at the cost of increasing complexity in the control-system structure.

1-5. Process Knowledge

The need for inter-loop coordination adds another dimension to the system-design problem. The number of possible coordinated configurations becomes a multiple of the factorial number for single-loop configurations. The probability of selecting the optimum configuration diminishes rapidly as the number of variables increases. Furthermore, the likelihood is not much greater of changing an inadequate configuration into a successful system by a simple exchange of connections, except in the 2×2 case.

As a result, control-system design cannot follow a cut-and-try approach, or even rely on duplication of earlier designs, no matter how successful they might have been. Furthermore, there are some processes that cannot be satisfactorily controlled by a single configuration, but rather require a distinct configuration for each operating regime.

The development of a suitable control-system structure begins with the formulation of a process model. We must know how the process responds to inputs before we can close loops through controllers. Steady-state relationships are the most important because they determine the fundamental stability modes of the process. Furthermore, they are relatively easy to obtain using material and energy balances and thermodynamic laws. However, dynamic characteristics are also important, in that they can produce responses that a steady-state model does not predict, capable of even destroying stability, or in other situations permitting improvements in structure. Dynamic information is much more difficult to obtain than steady-state equations, and here experience is the best teacher. Computer simulations can be helpful, but they also can lead to the wrong conclusions when based on invalid assumptions or when significant properties are unknowingly omitted.

Most control theory is based on linear representations of input-output relationships. Linear models are useful for illustrating dynamic behavior in response to small perturbations about a known operating point. However, it is a mistake to base the design of a control system on a linear process model. Most processes are nonlinear by nature, and cannot be adequately controlled over a broad range of production-rate or ambient variations using networks of linear components. While tests for process interaction may be made by differentiating input-output relationships, the compensation of interaction through feed-forward and decoupling functions is best realized by solving the fundamental nonlinear relationships themselves, rather than by using their derivatives.

The functions which are coordinated within the control-system structure must reflect the process relationships if the system is to be truly effective in maximizing process performance. Then, knowledge of the process is a prerequisite to the design of an effective control system. This is true not only for genera of processes but also for each specific unit. For example, general knowledge of distillation will help an engineer apply a control system to a distillation column, but it is not enough. He must also know the characteristics of the particular column to be controlled, its limitations, and most importantly, its objectives.

Each distillation column is sufficiently unique that its control system also will be unique, even if the nominal function of the column (e.g. the separation of methanol and water) is the same as that of another in a different location or constructed at a different time. This is especially true during a period of developing technology and of resource scarcities. Emphasis and values being in a state of flux brings about changes in objectives and methods, calling for continuing reassessment of the problem. Distillation is not the only operation requiring such individual attention, but is representative of most. Each dryer, boiler, compressor, refrigeration unit, evaporator, crystallizer, and particularly reactor, has its own peculiarities which must be understood by the control engineer if he is to control it successfully.

1-6. Developing a Structure

The simplest structure for a process having n controlled and n manipulated variables, would consist of n independent feedback-control loops. Each of the controllers would convert changes in a process output into the manipulation of a process input.

It is important to recognize the distinction in name and range between the input and output signals of the controller. Each controlled variable will have a controller assigned to it, whose input signal has the name, range, and units of that controlled variable. Occasionally, the designer may determine that the variable he wants to control is not one of the available measured variables, but a function of one or several measurements. Nonetheless, the name of the controller is the name of its controlled-variable input.

The output of the controller is whatever the system designer wants it to be. He should search among the manipulated variables for the one which uniquely influences the variable he wants to control. Once he has connected the controller to that manipulated variable, the output signal takes its name, range, and units. He will usually find that a single manipulated variable does not uniquely determine his controlled variable. He will have to decide which, among those that are available, has the most influence. Even then, there may be no clear choice. He should then look for some function of two or more variables that uniquely determines the controlled variable. He may assign that function to the controller output, and proceed to assemble a mathematical structure relating that designated output to the variables which take part in the calculation.

To illustrate the procedure consider this example. In assigning manipulated variables to the controllers for the boiler in Fig. 1-3, a designer might proceed as follows. Steam pressure is determined principally by steam flow and fuel flow. But steam flow is determined by the users and so is not available for manipulation. The logical alternative would be to manipulate fuel flow for steam-pressure control. The controller output would then represent fuel flow.

He could continue in the same way selecting outputs for his drum-level, draft, and flue-gas controllers, but conflicts will begin to appear: Flue-gas oxygen content is affected equally by fuel and air flow; furnace draft is affected equally by inlet and exhaust dampers. If he simply configures five single loops he can expect severe interactions which will create stability problems and allow disturbances to propagate from loop to loop.

At this point, he should break away from the given list of manipulated variables and consider appropriate combinations of them. For example, flue-gas composition can be seen to be determined uniquely by the fuel-air ratio. In this case, the output of the flue-gas composition controller should be the fuel-air ratio. It cannot then be connected to either single manipulated variable, but must coordinate the two. If fuel flow is already selected for steam-pressure control, then air flow must be manipulated in ratio to fuel flow, with the ratio manipulated by the flue-gas composition controller. Figure 1-4 compares this type of coordination with independent single loops for these two pairs of variables.

There are, of course, many other considerations which will modify the structure as described in later units; nonetheless, this is how it begins. Note particularly the presence of cascade flow controllers for both fuel and air. They are not essential to achieve the type of coordination described above, but improve its accuracy and responsiveness. The flows are then both manipulated *and* controlled variables—manipulated by the pressure and composition controller and controlled by the flow controllers. Because the flow loops are much faster than steam pressure and flue-gas composition, the cascade configuration gives a definite advantage—this is not always the case. The secondary (inner) loop should be at least four times as fast as the primary (outer) loop for cascade control to present a dynamic advantage.

As more loops are added, more interconnections should be made, gradually building an integrated structure. Signals are combined by means of mathematical operations which are often nonlinear —the multiplier in Fig. 1-4 is a case in point. The boiler control system itself becomes quite complex, and, for reasons which will be explained in Unit 9, may not include the elementary structure shown in Fig. 1-4, although it features the same function.

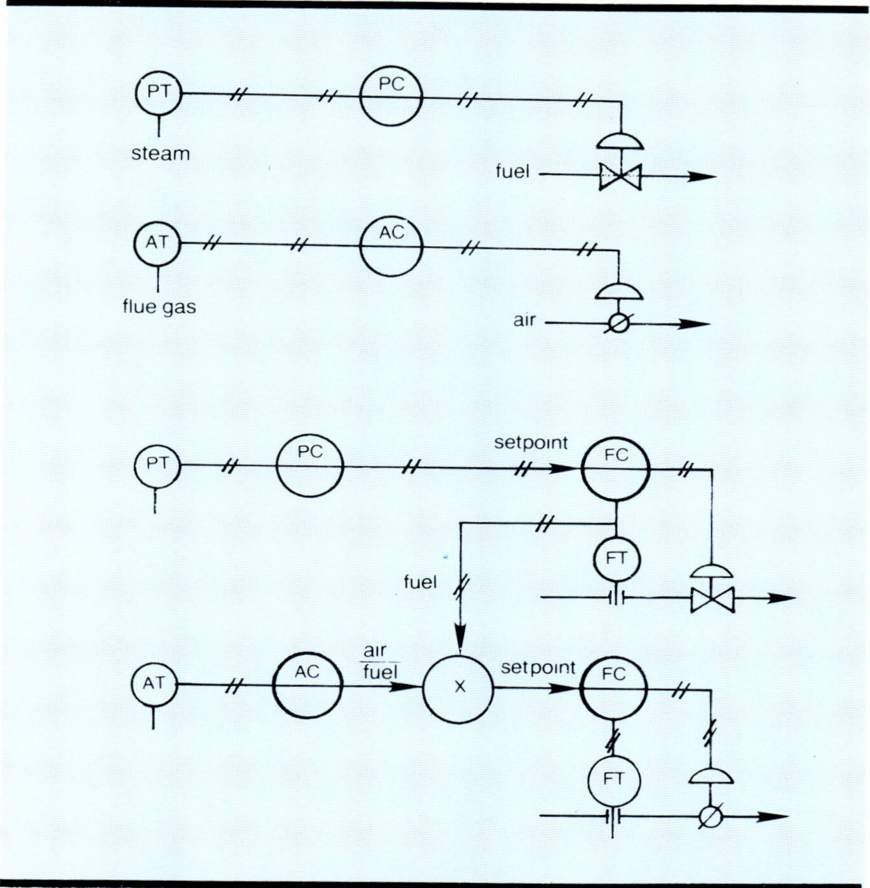

Fig. 1-4. Coordinated controls (below) reduce interactions permitted by independent single loops (above).

At this point, the introduction is complete. Later units develop each of the foregoing points in detail, using real process applications to illustrate the concepts. To evaluate your understanding of this unit, complete the following exercises.

Exercises

1-1. *Identify the types of controlled variables appearing in Fig. 1-2.*

1-2. *Is the interaction between the variables in Fig. 1-2 likely to be severe? What does it depend on?*

1-3. *What would be the disadvantages of applying two independent control loops to the process in Fig. 1-2?*

1-4. *There are more transmitted variables in Fig. 1-3 than there are valves and dampers. Explain the discrepancy.*

1-5. *Are there likely to be significantly more transmitters than valves and dampers in a typical multivariable control system? Why?*

1-6. *The system structure in Fig. 1-4 was developed intuitively. What are the limitations of this procedure?*

Unit 2: Production-Rate Controls

UNIT 2

Production-Rate Controls

This unit describes how to establish the production rate for a process and the conditions under which the rate must be limited. Startup and shutdown are also discussed.

Learning Objectives — When you have completed this unit, you should:

A. **Understand how production rate is established in common industrial processes.**

B. **Recognize the factors that limit production rate and how to cope with them.**

C. **Be able to devise systems and procedures for operating within constraints, and allow automatic startup and shutdown.**

2-1. Locating the Production-Controlling Variable

Every process has a mechanism or "handle" by which production rate is established. In an automobile that mechanism is the accelerator, through which speed is regulated by the driver. The speed is subject to various constraints such as speed limits, stop lights, and traffic.

Production in a modern industrial plant is not much different, except that there are fewer variations, and costly stops and starts are avoided insofar as possible. In some processes, the variable that sets production rate will be obvious to the engineer who is given the task of designing the control system. But this is not always the case.

Considerable inquiry into the objectives and characteristics of the process and an examination of operating procedures may be required to find the manipulated variable that has the primary influence over production rate.

Quite often, that variable may encounter constraints or "speed limits," which it may not exceed. A limit may be arbitrarily set within the control range, or it may be mechanically imposed, e.g., by a wide-open control valve. In the latter case, production rate is uncontrolled and limited. Recognize also that production entails the coordination of many streams, any one of which could encounter a limit; the control system must be sensitive to this, transferring production-rate control to the limiting stream. To maintain satisfactory performance in both limited and unlimited modes, the control engineer must locate all the principal streams to be coordinated, find the leader, and determine how to exchange roles when required.

Part of the difficulty in finding the production-controlling variable is that it may lie in another part of the plant, entirely beyond the scope of the unit whose controls are being designed. Then the unit under consideration is simply a slave imposed by the production-rate control, wherever it is.

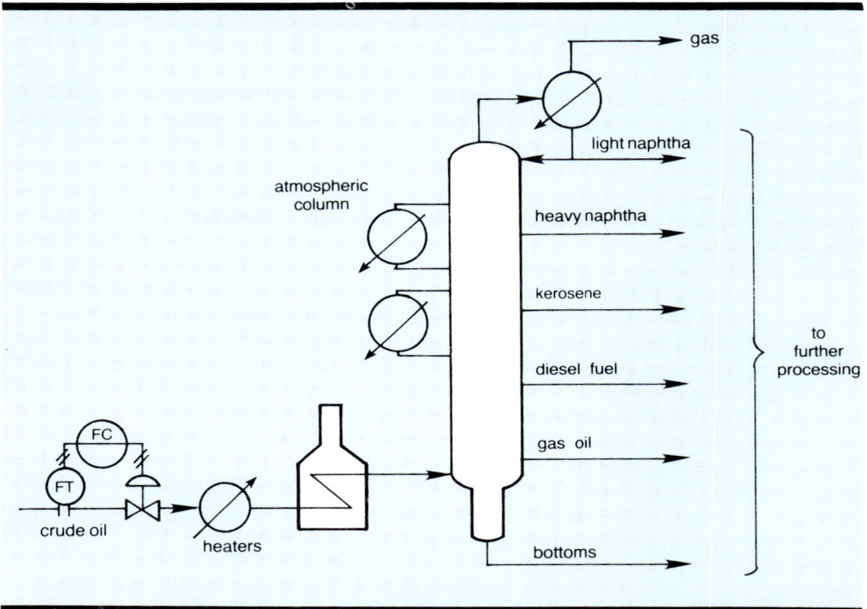

Fig. 2-1. The charge rate of crude oil determines the throughput of all downstream operations.

An important point to remember is that there is only *one* production-rate control active in a line at any time. In an oil refinery, it is the feed rate of crude oil to the atmospheric still as described in Fig. 2-1. The flow rates of all products leaving that column in turn set production for the facilities which process them. But their rates are entirely dependent on the crude charge rate, crude composition, controls on the still, and the efficiency of the still. Process units farther from the point of the production-rate control will experience ever-increasing variations in throughput resulting from all upstream disturbances, as well as weather changes. Surge tanks located between units can smooth some of these fluctuations dynamically, but have no effect on the steady-state throughput. Their roles and effectiveness are discussed under Inventory Controls in Unit 3.

2-2. Setting Supply and Demand

Most processes have their production rates set at the supply side. This is true for the crude-oil fractionation discussed above, and also for chemical plants. Because of increasing variability experienced by streams farther from the rate-control point, that point is usually located to favor the most critical operation in the plant.

In the chemical plant, that operation is the reactor. To optimize yield of product, minimize unwanted byproducts, corrosion, catalyst loss, etc., reaction conditions are controlled as carefully as possible. Therefore all ingredients are combined as accurately as measurable and controlled at a uniform flow rate, which usually determines reactor residence time and thus sets conversion. Under normal conditions, reactor outflows will be nearly constant, and stream compositions will vary only with such slowly developing events as the decay of catalyst. Production-rate changes are made only infrequently, to reflect demand for product or limitations of supply-side or product-side inventory.

There are some abnormal conditions developing from time to time, but most of these are handled in a controlled manner. Reactors are commonly taken out of service for cleaning or catalyst regeneration, and replaced by others. These operations can initiate sharp supply upsets in both rate and composition for downstream units, but they are operator-initiated and generally predictable in their effects. Disturbances of this nature do not have to cause difficulties downstream because they are amenable to feedforward compensation, as described in Unit 8. This is generally true of all supply-side upsets, but demand-side upsets can cause feedback problems which need to be carefully addressed.

Such facilities as boilers and refrigeration units have their production rates set by demand. The boiler in Fig. 1-3 is a case in point—steam flow was measured but uncontrolled, subject to the aggregate demand of all users. The same is true of power plants in general—the rate of electric-power production is set by usage, both in central stations and industrial generation facilities. Another dimension of the problem is that these units have no product inventory. They must respond instantly to demand variations if they are not to lose control over product quality.

The lack of product inventory is mitigated to some extent by increasing the size of the distribution system. As the number of users increases, the probability of a large, instantaneous change in demand is reduced.

Placing many users on a single distribution network requires multiple sources of supply as well. Thus, several boilers commonly share the demand for steam and electricity in an industrial powerhouse. The most efficient are usually "base-loaded," i.e., set at a constant production rate by supply-side controls. This is an economic measure which allows the less efficient to be turned down during periods of reduced demand, thereby saving fuel. However, the removal of some of the supply capacity from the role of meeting demand variations increases the rate changes imposed on the responding units. It also increases the likelihood of failing to meet demand as constraints are reached. In this respect, base-loading carries a risk. But, because it is practiced for economic reasons, it can be automated by economic control loops to maximize performance without the usual risk. The solution to this problem is developed in Unit 6.

In most process plants, steam is provided at several pressures. The highest pressure is delivered by the power boilers to steam turbines, which may generate electric power or drive rotating machinery directly. Steam at progressively lower pressures is extracted from or exhausted by these turbines. Changes in demand for electric power or shaft work are met by the turbine controls and passed back to the boilers. Changes in steam demand at all levels are met by regulating valves and, to avoid upsetting turbines, are also passed back to the boilers. Therefore the boilers must respond not to a single disturbance, but to an aggregate of all disturbances arising not only in the steam-distribution system but also in the plant electrical network.

Boiler steam flow is used as an index of demand, but it also contains supply-side information. An unrequested change in firing rate will affect steam flow in absence of a change in demand. Hence the flow signal by itself cannot be used in feedforward manipulation of firing rate or a positive-feedback loop is formed. Fortunately, this problem is easily resolved by compensating the flow measurement with steam pressure, as described in Unit 11.

Occasionally production rate is set somewhere between the feed point and the product discharge, owing to the presence of a critical intermediate operation. The usual culprit is a chemical reactor located downstream of a separation unit, such as a distillation column, as shown in Fig. 2-2. This location of the production-rate control can cause real problems for upstream operations, most of which respond poorly to demand-side upsets. Inventory controls, such as the level controller in Fig. 2-2, must cascade backward instead of forward. While this is accomplished easily for simple storage vessels, the many trays between the feed and bottom of a column introduce delays, severely degrading level control. Often the only solution is to enlarge the inventory of reactor feed so that its time constant greatly exceeds the delays in the level-control loop. The same consideration must be given to all upstream units. Nonetheless, fixing a steady production rate to a reactor is the first order of importance—inventory control problems always can be solved as they arise.

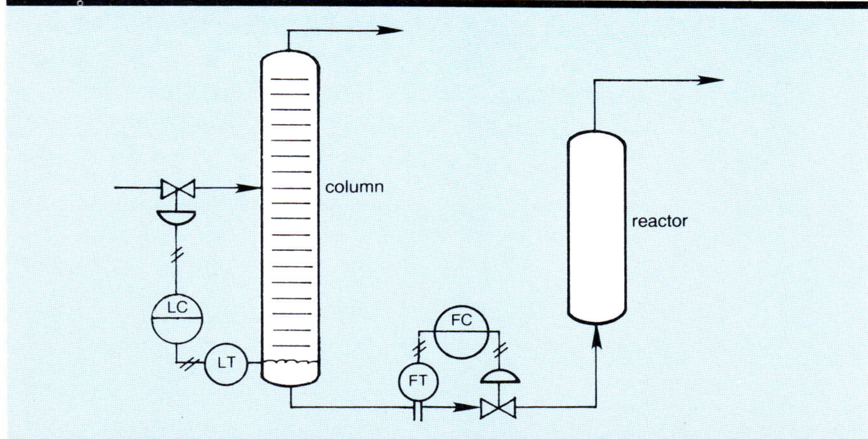

Fig. 2-2. A reactor following other processing steps forces upstream units to operate in a demand mode.

2-3. Coordinating Feed Streams

In a separations unit, a single stream may set production rate. Even then, services such as steam or fuel need to be matched to the rate of feed if product quality is to be maintained. However, in controlling reactors and product-blending systems, the need for stream coordination is more acute. Momentary excursions in the feed mixture to a reactor, like the fuel-air mixture to a furnace, not only could reduce process efficiency, but also damage equipment and endanger lives. Many of the direct-synthesis routes for oxygenated hydrocarbons, for example, involve mixing hydrocarbons like ethylene with oxygen near an explosive limit. These streams must be blended with extreme accuracy and frequently over a wide range of flow rates. Performance may be even more demanding where the rate-controlling flow is not fixed but is a function of upstream operations. This is the case for a sulfur tail-gas plant, in which air must be accurately proportioned to the amount of hydrogen sulfide in a stream whose flow and composition vary continuously. A similar requirement is placed on in-line product-blending systems, such as those that deliver gasoline directly into a pipeline or a tank truck. The absence of a large downstream capacity means that the composition of the mixture must be controlled continuously by exact proportioning of the various ingredients.

There are several ways of bringing this about. If there is one base stock that is always present in the majority, it may be controlled either individually or manipulated for total flow control, as shown in Fig. 2-3. This arrangement also is viable when the flow of base stock is uncontrolled. It is not good practice to set the additive streams in ratio to total flow as this forms a positive-feedback loop, reducing stability and increasing settling time.

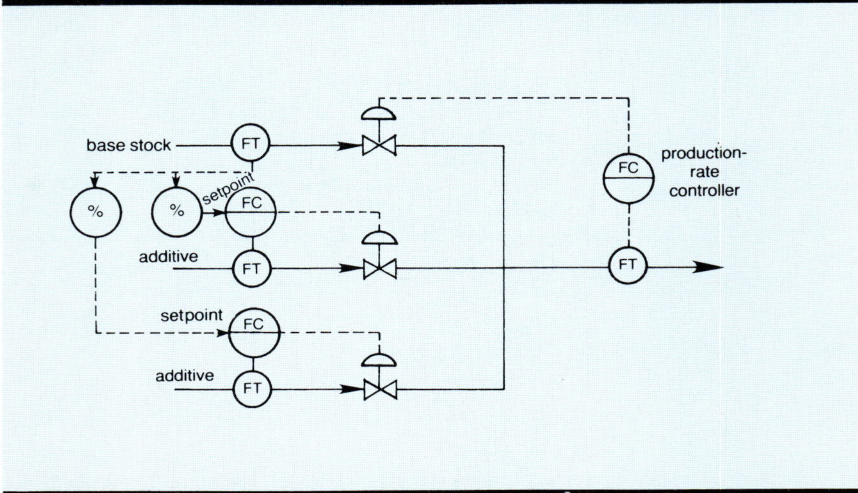

Fig. 2-3. Blend composition is regulated by setting additive flows in ratio to that of the base stock.

Where a system must prepare a variety of products, possibly no single stream can be designated as a base stock for all. Also, very wide ratio ranges are to be expected. Then the production rate is set at a master station, which in turn sets *all* streams as percentages of its output, as shown in Fig. 2-4. This system can be made to respond to demand by replacing the production-rate setter with whatever controller establishes the demand for product.

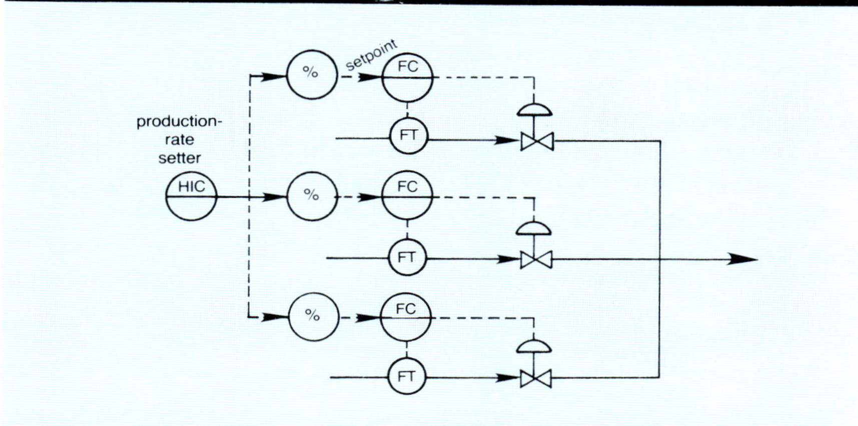

Fig. 2-4. Blend composition here is regulated by setting each flow as a percentage of the total set production rate.

Sometimes a total-flow measurement and controller are used in place of the production-rate setter in this type of installation. This practice can give unsatisfactory performance in that it has a flow controller setting other flow controllers in cascade, with no capacity or time lag between them. The total-flow controller then needs a very wide proportional band and relatively long integral time for stability.

2-4. Production Constraints

There always will be real constraints placed on production. They are more likely to be encountered at high rates than at low, and when a plant must be operated at maximum production, control at one or another constraint is a normal occurrence.

If the constraint is imposed directly on the production-rate variable, resulting in a controlled or uncontrolled reduction in flow, all streams set in ratio to its flow will follow. Then the proper proportioning of ingredients will be maintained at the reduced production rate. This is the situation that would prevail if the supply of base stock in Fig. 2-3 became limited below what was requested by the production-rate controller, perhaps by a reduction in supply pressure. Then total flow would fall below setpoint, with a flow controller responding by fully opening the base-stock valve.

A similar cutback would follow high pressure on the discharge side. Alternately, a level controller on product-tank inventory or on feed-tank inventory could override the base-stock valve, as shown in Fig. 2-5. Shown is a low selector comparing the three controller outputs, sending the lowest to the valve. When tank levels are normal, their controllers will try to increase flow, but the production rate controller will limit it. Excursion of either tank level beyond its setpoint in the unsafe direction will cause flow to decrease immediately. Figure 2-5 shows the selected output being fed back to the integral mode of all controllers. This is necessary for transfer of control exactly at setpoint, as described in Ref. 1.

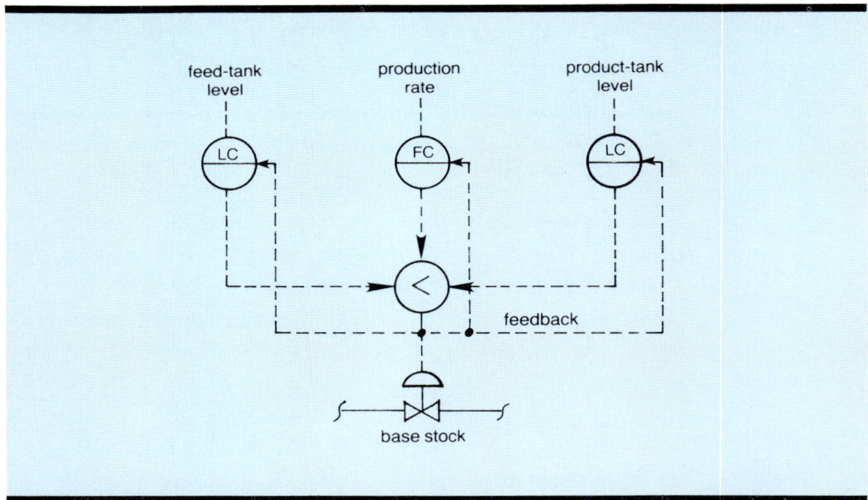

Fig. 2-5. In case of a low feed-tank level or a high product-tank level, the appropriate level controller can reduce base-stock flow through the action of the low selector.

For the system where all streams are set in ratio to the production-rate demand in Fig. 2-4, overrrides from tank levels, or any other source, could be applied to a low selector acting on the output of the rate-setting station.

There are many more constraints that conceivably could be applied to production-rate control, but their implementation is not always the same as shown in Fig. 2-5. Consideration needs to be given to the speed of response of the constrained variable relative to that of the controlled variable. For the system in Fig. 2-5, upstream and downstream tank levels presumably respond directly to the flow of base stock, and so the imposition of constraints is straightforward. But occasionally the constrained variable is affected only indirectly, through the action of intermediate control loops, complicating the issue. Then a more complex structure is required to protect the constrained variables, involving those intermediate loops in its operations. These systems are discussed in Unit 10.

2-5. Pacing Systems

The effects of limits on other *manipulated* variables must also be considered as constraints on production rate. Suppose that the supply of one of the additives to the blend in Figs. 2-3 or 2-4 became restricted, such that its flow controller was unable to deliver the set flow even with its valve fully open. This may not appear as an abnormal tank level, and, in fact, may be due to any of several obscure factors, such as a clogged strainer in the additive line.

Whatever the reason, failure to deliver enough of the additive will cause the blend to deviate from the desired composition. While the problem could be detected by an on-stream analyzer, some off-specification material will be produced before corrective action can take effect. Also, the type of corrective action required is not clear in that there may be many additives, any one of which could fail. Furthermore, analyzers may not be available for all components of the blend.

A preferred solution to the problem is to base action on the *impending* loss of flow control, as valves approach their full-open position. In this way, production can be curtailed *before* off-specification product is made. Systems which allow any of the manipulated variables to curtail production are called "pacing" systems. Any valve may set production rate, based on its approach to a fully open position. Figure 2-6 shows a high-signal selector (>) sending the position of the most-open valve to a valve-position controller (VPC). Its function is to reduce the flow of base stock to keep all valves from exceeding 90% open (or whatever setpoint is chosen as the upper limit).

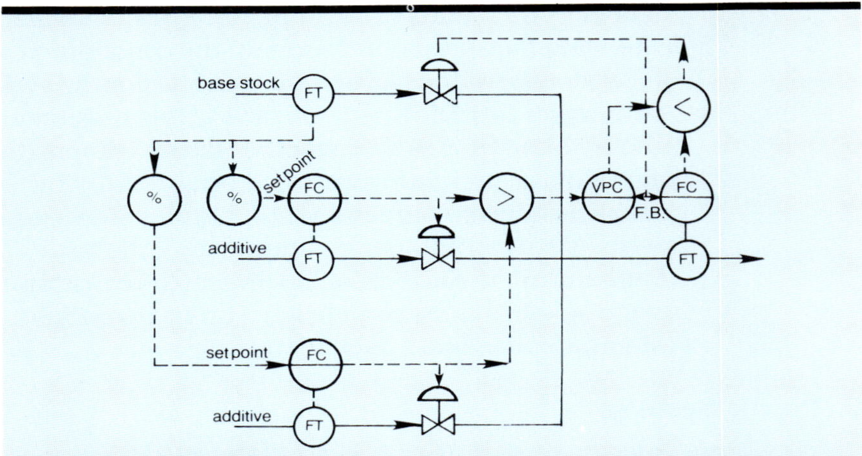

Fig. 2-6. The valve-position controller will reduce the flow of base stock to keep the most-open valve from exceeding 90% open.

In the blending system, the subsequent reduction in flow of base stock is fed directly through the ratio stations to the additive flow controllers, closing the valve-position loop. The fast response of this loop then can be expected to control valve position quite tightly—response will be much superior to any system that uses an analyzer for constraint control.

For the blending system in Fig. 2-4, the valve-position controller would look at the most open of *all* valves (there being no base stock) and override the production-rate setter as described above.

The technique of pacing is practiced principally in blending systems, but is not limited to them. Any of the valves involved in controlling variables affecting production can participate in a pacing system. A common problem in distillation is the inability to provide enough steam to a column reboiler during conditions of cold weather, high production rate, unusual feed condition, or accumulation of noncondensible gases in the reboiler. Any of these conditions could cause the steam valve to be driven fully open, with subsequent loss of control of its assigned variable. A valve-position controller acting on the signal to the steam valve could then be used to reduce feed rate when full opening was approached.

Care must be exercised in applying these valve-position controllers, however, in that they cannot operate faster than the loops they depend on for complete closure. In the blending system of Fig. 2-6, the action of the valve-position controller caused the flow of base stock to change rapidly. This change was fed *forward* by the ratio station to the additive flow controller, which promptly responded by moving the valve. Without the feedforward ratio function, the valve-position loop would be open. In this open-loop mode, the VPC cannot regulate production rate, but it can be used for startup and shutdown. This function is examined under Sec. 2-6.

If steam flow to the distillation column were set in ratio to feed rate, then a VPC on the steam valve could act in the same manner as that in the blending system. However, if steam were manipulated by an independent, single-loop temperature controller, valve-position control would have to be much slower. A reduction in feed rate would have to cause a rise in temperature, which the temperature controller would have to act upon, before the valve position were to change. The ensuing delay could result in the production of considerable off-specification material. Reducing the VPC setpoint would help curtail feed rate earlier and thereby reduce that likelihood, but at the cost of a lower production-rate limit. In this context, feedforward becomes an important ingredient in pacing systems.

2-6. Startup and Shutdown

Startup of a continuous production facility requires the careful coordination of many variables whose flow patterns are often different from what is provided under normal conditions. Certain constraints may limit the rate-of-approach to production conditions that are not significant during steady-state operation. The rate of temperature rise in heat-transfer equipment and turbines, for example, must be limited to avoid deformation of expanding components.

Units which, under normal circumstances, are demand-following, have to start operation by manual or automatic ramping of production rate. Transition to demand-mode operation should be automatic, and provision should be made for transfer to a shutdown mode when necessary.

Consider the boiler-turbine facility in Fig. 2-7. Rate of temperature rise must be limited at both boiler and turbine—in this figure, both are accomplished by operator-set stations (HIC). Pressure controllers override to hold boiler pressure between minimum and maximum limits to reduce the risk of operator error in matching the supply and demand. A minimum-flow controller is also provided to protect heat-transfer surfaces from overheating during startup.

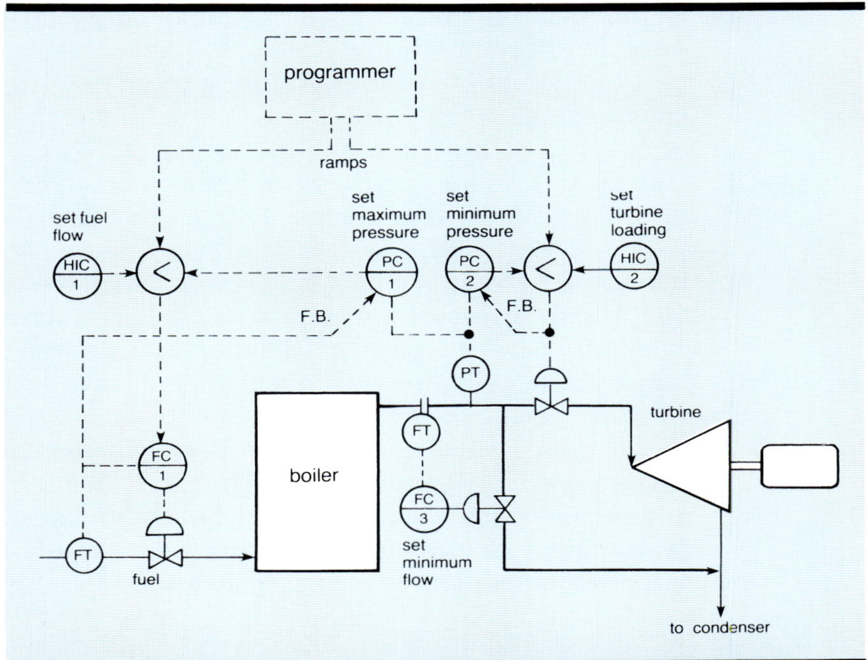

Fig. 2-7. This system allows the operator to set fuel flow and turbine loading independently during startup.

During initial conditions, both HICs would be set at zero. With no steam-generation, the minimum-flow controller would open the turbine bypass. The operator then would ignite burners at minimum fuel flow. The flow controller would bypass any steam produced around the turbine. Turbine loading could not begin until a certain minimum steam pressure were reached, regardless of the setting of HIC-2. If HIC-2 is set high, PC-2 will begin controlling when the firing rate has generated enough steam in excess of FC-3 setpoint, thereby causing boiler pressure to rise to PC-2 setpoint.

In actual practice, however, the operator may increase HIC-1 and HIC-2 together, watching temperatures closely. Steam production must be higher than turbine loading if boiler pressure is to rise. As turbine loading is increased, FC-3 will close the turbine bypass valve automatically.

Boiler pressure can rise only to the setpoint of PC-1, at which point PC-1 will assume regulation of firing rate. Then the plant is in the familiar boiler-following mode of operation, with load set by the turbine. Then HIC-1 can be raised to the maximum firing limit desired. In an overload, boiler pressure can fall only to the setpoint of PC-2, at which point the boiler becomes base-loaded at the setting of HIC-1. The unit may be shut down by reducing the setting of either HIC-1 or 2 or both, as desired. Because both signal selectors are low selectors, production rate is determined by the lower setting of the two HIC stations.

To program the rate of startup, a ramp generator could input to both low selectors, in which case the HICs would be set at fixed high-limit settings. Both firing rate and turbine loading would then rise until either a PC or HIC were selected; the ramp would stop at that point. Shutdown could be accomplished by reversing the ramp.

Alternately, the programmer could be replaced by differential temperature controllers connected to pairs of critical points in boiler and turbine. The temperature controllers would regulate temperature rise as determined by a specified set difference between said critical pairs of measurements. Then the rate of increase of loading would be the maximum permitted by temperature constraints. Again, HICs and PCs would limit maximum loading.

A change of operating modes signaled by opening or closure of certain valves may also be used to initiate startup action. The system shown in Fig. 2-8 allows completely automatic startup from low fire to full load. The rate of low fire is set by HIC-3. When minimum operating pressure is reached, PC-2 will increase turbine loading at a rate limited by the temperature-difference controller DTC-2. As long as the turbine bypass valve remains open, the boiler is held at low fire by the low output from VPC-3. However, when turbine loading has increased to the point where FC-3 closes the bypass valve, VPC-3 will begin to ramp its output up from the low-fire value. The ramp will continue upward to the limit set by either PC-1 or HIC-1, because the valve-position loop is open.

A turbine trip, which opens the bypass valve, will cause VPC-3 to automatically ramp the fuel flow back down to low fire.

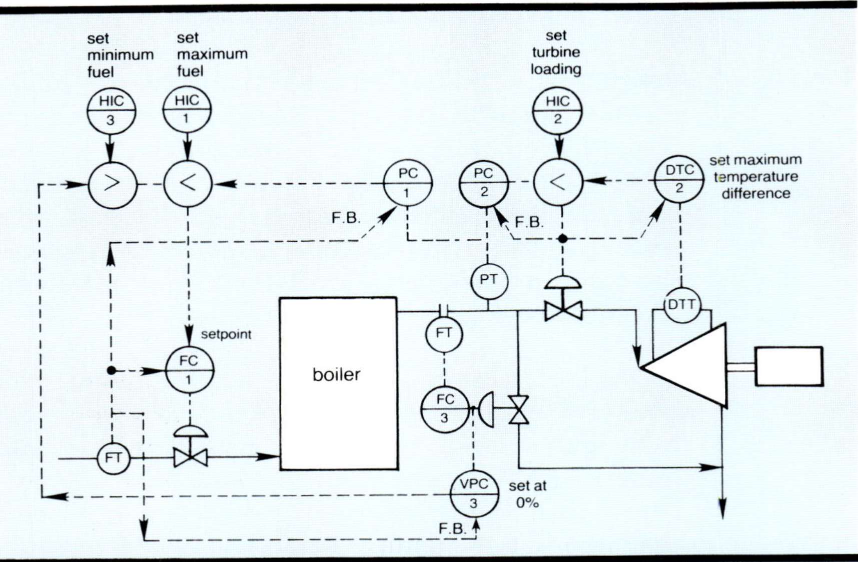

Fig. 2-8. Here, turbine loading is limited by the temperature-difference controller, and ramping of fuel flow is initiated by closure of the bypass valve.

Hopefully, the use of examples to illustrate the functioning of production-rate controls has identified the concepts well enough to enable the reader to apply them to similar processes. To test your skills in these procedures, work the following exercises.

Exercises

2-1. *Many of the products from the crude-oil still in Fig. 2-1 are fed to reactors under constant flow control. How is their material balance satisfied?*

2-2. *Why is dynamic response important for demand-following processes like boilers? How can it be improved?*

2-3. *How do the settings of the ratio station in Fig. 2-4 differ from those in Fig. 2-3?*

2-4. *If the production-rate setter in Fig. 2-4 were replaced by a measurement of total flow, what would happen?*

2.5 *Suppose a product-quality analyzer and controller were used to adjust the ratio (%) station in Fig. 2-6. Would the VPC still be necessary? Why?*

2-6. *Add turbine-trip logic to the bypass valve in Fig. 2-8. Will the boiler be ramped to low fire faster than it was ramped to turbine loading? Why?*

References

[1]Shinskey, F.G. *Process-Control Systems*, 2nd Ed. New York: McGraw-Hill Book Co., 1979, pp. 153, 154.

Unit 3:
Inventory Controls

UNIT 3

Inventory Controls

This unit describes how to control inventories of material and energy at locations scattered throughout a process and to coordinate inventory with production rate and its constraints.

Learning Objectives — When you have completed this unit, you should:

A. Be able to place inventory controls throughout a plant flowsheet.

B. Recognize the need and technique for variable inventory control.

C. Be able to close mass balances around chemical reactors, both with and without recycle streams.

3-1. Material and Energy Balances

Both mass and energy are conserved in every process. They therefore must be removed from a process at exactly the same rate they are delivered to it, if a steady state is to be maintained. Some processes discharge material or energy directly into the environment, and these may then float on the environment. For example, there are atmospheric distillation columns which have no pressure controls because they are vented to atmosphere. (This is not necessarily efficient or desirable, but it is nonetheless practiced.) Similarly, an air compressor that draws suction from the atmosphere needs no suction-pressure controller.

In most cases, however, material and energy are flowing from a source through a series of process units before reaching the environment. Each of these units needs its material and energy balances closed. If inflow and outflow are not matched, there must be accumulation. Then the difference between inflow and outflow is sensed and controlled in the accumulation of inventory.

The capacity for storage of material and energy varies widely. A liquid-filled system has almost no capacity to absorb differences between inflow and outflow because liquids are almost incompressible. An imbalance between inflow and outflow will cause pressure to change instantaneously—therefore liquid-pressure regulators are needed to balance these units. Open vessels, on the other hand, can accommodate substantial variations in liquid inventory, which is sensed as the liquid level within them. Gas is much more compressible, and so gas-filled systems show less pressure change on flow imbalances than liquid-filled systems.

Within the overall material balances there are also individual component balances which have to be satisfied. If concentrations of components in a mixture are constant, then regulating the overall material balance will also regulate that of each component. But in many processes, especially those involving chemical reactions, concentrations change, and some species may accumulate. Recycle streams can complicate the issue in that they form positive-feedback loops, although not necessarily for all components. Composition analyzers are essential to allow control of component inventories in reaction systems.

Capacity to store energy also varies widely. Energy is stored in the heat capacity of solids, liquids, gases, and in the vessels that contain them, being sensed as changing temperature. But it is also stored as latent heats of fusion and vaporization, which may not be accompanied by temperature change. These operations feature an exchange of inventory between the solid and liquid, and liquid and gaseous phases; hence, energy inventory may then be sensed as liquid level and pressure.

3-2. Self-Regulation

Self-regulation is the property of being able to reach a steady state without the application of controls. Self-regulating processes have either their inflow or outflow or both affected by their inventory of material or energy.

Temperature is self-regulated in many processes: An increase in heat input raises the temperature of the leaving stream, thereby carrying away more heat. Eventually the temperature will rise to the point where a new steady state is reached. This is also common with pressure: Increasing inflow raises pressure which causes an increase in outflow, driving the process toward a new steady state.

Liquid level can behave in a similar fashion but rarely does. Raising the level in a tank can increase outflow or decrease inflow significantly only if the tank is open to the atmosphere and its feed or discharge line is similarly open. In process plants, this is rarely true. Pressure drops across inlet and outlet valves are usually much greater than the change in liquid head between low- and high-inventory levels. Then *any* difference between inflow and outflow will cause inventory to either rise or fall—a steady state can be achieved only by applying a controller.

Nonself-regulating processes pose some special control problems. They cannot be left unattended in manual control because eventually the controlled variable will surpass allowable limits. An integrating-only controller cannot be used on them or sustained oscillations will result (1). If any controller with integrating action is used, and there is measurable dead band in the loop (which is common to control valves without positioners), a limit cycle will result (2).

In addition to the pressure-drop relationship cited above, there are certain other process characteristics which eliminate self-regulation. Positive-displacement pumps deliver the same flow regardless of suction or discharge head and so inhibit self-regulation. Automatic flow controllers accomplish the same function. Recycle streams can also destroy self-regulation by returning materials already discharged by the process.

Endothermic reactors consume heat and are therefore highly self-regulating: Increasing reaction rate absorbs more heat, thereby lowering temperature and reducing reaction rate.

Exothermic reactors vary from marginally self-regulating to having no steady-state stability whatever, depending on the characteristics of their heat-removal systems. Increasing reaction rate generates more heat, thereby raising temperature, which increases reaction rate. To counter this positive feedback, a cooling system is needed which is highly self-regulating, such as a boiling-water bath, which needs no rise in temperature to carry away more heat (3).

3-3. Constant Inventory Control

In most processes, constant inventory control is required. This is important not only for maintaining the process in a steady state, but also to maintain its efficiency constant and to protect equipment from damage.

Consider, for example, the liquid level in the steam drum of a boiler. If the level is allowed to rise too high, water may be carried out of the boiler and possibly damage turbines, etc. If it should fall too low, circulation may be affected in the boiler tubes, resulting in overheating and damage. Similarly, steam pressure is usually held constant to maintain constant efficiency of both boilers and turbines.

Liquid-level control is important in all boiling vessels to keep heat-transfer surface covered and thereby maximize heat transfer and minimize fouling. It is also essential upstream of turbines and compressors to keep liquid out of the machinery.

In controlling chemical reactions, it is necessary to match ingredients precisely. If any component is in excess, some will be unconverted and may be lost if not recovered for recycle. It could also be corrosive, damaging equipment if allowed to accumulate. Usually, reactors are controlled for zero excess, but in some cases, fewer side reactions occur if a particular component is held in an excess, but constant, concentration.

Constant inventory control requires a very high controller gain, achieved either by a narrow proportional band or integral action. Feedforward is occasionally used when sufficient controller gain cannot be applied due to stability limitations.

Regardless of the mechanism used, the act of holding inventory constant necessarily passes along all incoming upsets directly to the outputs of the process. In some instances, this is unavoidable and is a price worth paying for the control of a critical inventory. But often it causes more problems than it solves. It is not uncommon to see liquid levels in all reboilers in a series of distillation columns cycling at the same period. A disturbance or cycling in the first is passed through the complete series because they are all on constant inventory control. Furthermore, if all are of the same relative size and geometry, their closed loops will tend to exhibit a high sensitivity in the same dynamic range. Then a disturbance developing in an upstream unit may actually be reinforced by downstream controllers.

In trying to hold inventory constant, a controller essentially eliminates any capacity the process has for absorbing fluctuations in throughput. A nonlinear controller of the form shown in Fig. 3-1 is useful here, for attenuating small disturbances with a low gain while taking vigorous action in the presence of larger upsets (4).

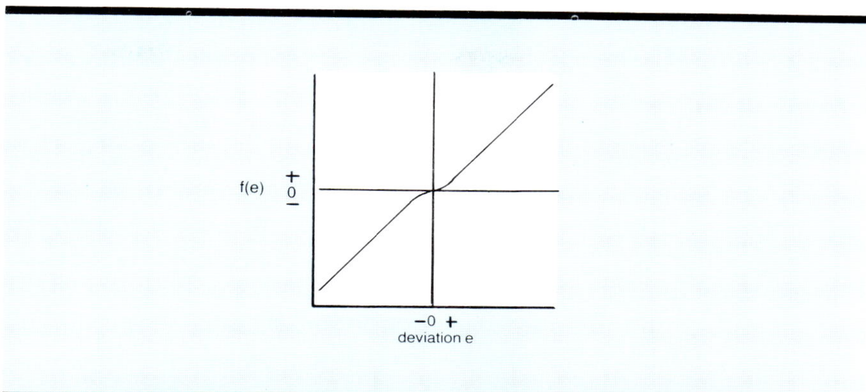

Fig. 3-1. This type of nonlinear control function filters noise and small disturbances without restricting control action in response to major upsets.

There can be no dead zone in the nonlinear controller, however. Unless there is some proportional gain at setpoint, lack of self-regulation in the process will cause liquid level to float through the dead zone and rest at either edge of it. Then the nonlinear filtering action will be only half as effective as if the level comes to rest at setpoint.

3-4. Floating Inventory Control

When there is no need to control inventory precisely to protect equipment, but rather the capacity is provided for smoothing, a different approach is needed. Inventory should be allowed to float with load in such a way that disturbances are attenuated as much as possible. If the *feed* to a storage vessel is the disturbing variable then liquid level should be allowed to rise with production rate. Then, at maximum rate, the vessel will contain the maximum amount of material to maintain a high outflow during subsequent reductions in inflow. If the *demand* from a tank is the disturbing variable, its inventory should be minimum when production is maximum, so that it will have all its capacity available to absorb feed when the demand is decreased.

These features are easily implemented using proportional-only control. A proportional liquid-level controller can manipulate the setpoint of the outflow controller. The level controller would be calibrated so as to develop zero flow setpoint at the minimum desirable level and maximum flow at the maximum allowable level. The flow measurement should be linear to keep the time constant of the tank constant. If the level controller were to drive a valve directly, the nonlinear installed characteristic of the valve would alter the vessel time constant as a function of stem position. When manipulating inflow, the controller action is reversed.

This same concept can be applied to maintaining a constant residence time in a chemical reactor for various conditions of production rate. In this case, the proportional band of the level controller determines the reactor residence time. If desired, the proportional controller could be replaced with a divider, as shown in Fig. 3-2. Vessel volume, as measured by liquid level, would enter the numerator while the denominator could be set as the desired volume-to-flow ratio, or residence time.

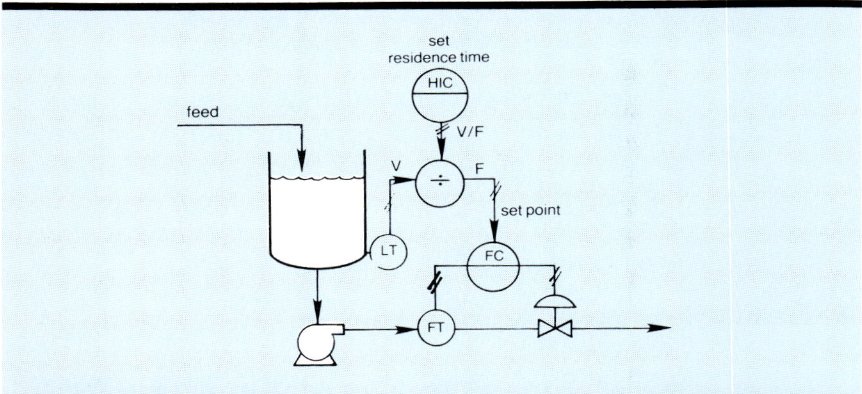

Fig. 3-2. Outflow is set directly proportional to liquid level to regulate the vessel's residence time.

The nonlinear function of Fig. 3-1 can also be used to provide smoothing over the entire range of tank level. The gain of the low-gain region can be set much less than 1.0, with the high-gain segments used to keep the tank from overflowing or emptying. If the integral mode is used also, liquid level will tend to return to the setpoint in every steady state, regardless of flow. This may be desirable in some instances, but it will not regulate residence time.

3-5. Endpoint Control

Endpoint control is vital to the successful operation of most chemical reactors. There are two basic methods of feeding reactors:

1. Ingredients are fed in stoichiometric proportions (i.e., in the proportions in which they react). Here, endpoint control is essential.

2. One ingredient is fed in substantial excess to insure complete conversion of the other(s). The excess must then be recovered and recycled. Endpoint control of the reactor is unnecessary, but it may be required of the recycle stream.

It is not possible to control the flow of individual reagents so accurately that no excess remains unreacted. In some cases, small amounts of unused reagent are not detrimental, but more often they cause problems. Unused reagent raises the cost of operating the plant and often brings corrosion or pollution. Acid-base reactions are a case in point. Unless acid and base are proportioned with extreme precision, effluent pH will be either too high or too low, attacking pipe and fittings and bringing death to any living organisms downstream.

A pH control system for a production reactor is shown in Fig. 3-3. Acid and base reagents are flow-controlled in a ratio that is adjusted by the pH controller. Errors in the flow measurements and variations in reagent strengths will be detected as changes in effluent pH. The pH controller will react to these disturbances by readjusting the reagent ratio to restore the stoichiometric balance.

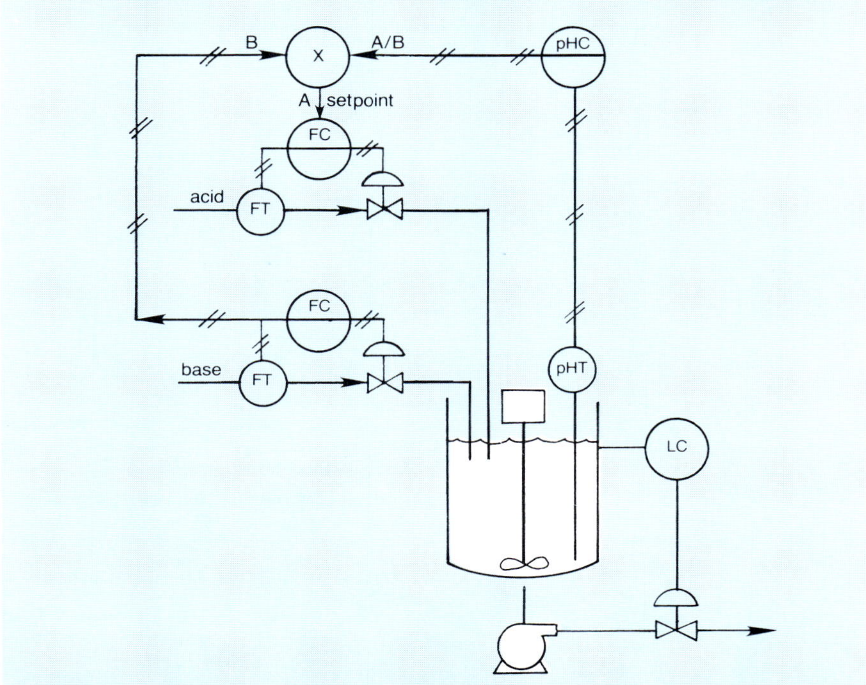

Fig. 3-3. A nonlinear pH controller manipulates the ratio of acid to base to control reaction endpoint.

In most reactions of this type, pH changes quite markedly with small differences between reagent flow rates and so is a very sensitive index of reaction endpoint. Sensitivity tends to decrease with larger differences, giving an overall relationship between pH and reagent flow which is highly nonlinear. The nonlinear control function described in Fig. 3-1 has been used quite successfully for compensating the nonlinear characteristic of many pH control loops.

If there is no phase change in the reaction, an analyzer of some sort is nearly always required to determine the endpoint. This is true when gases are reacted to form a gaseous product or when liquids are reacted to form a liquid product. But when the reactants are in two different phases, one of which disappears, endpoint can be controlled by liquid level or pressure.

Figure 3-4 shows how the pressure of gas above a liquid-phase reaction represents accumulation of excess gaseous reagent. Increasing the flow of liquid reagent will increase its concentration in the reactor, thereby consuming more of the gas present. This will cause pressure to fall until the pressure controller raises gas flow to reach the new steady state.

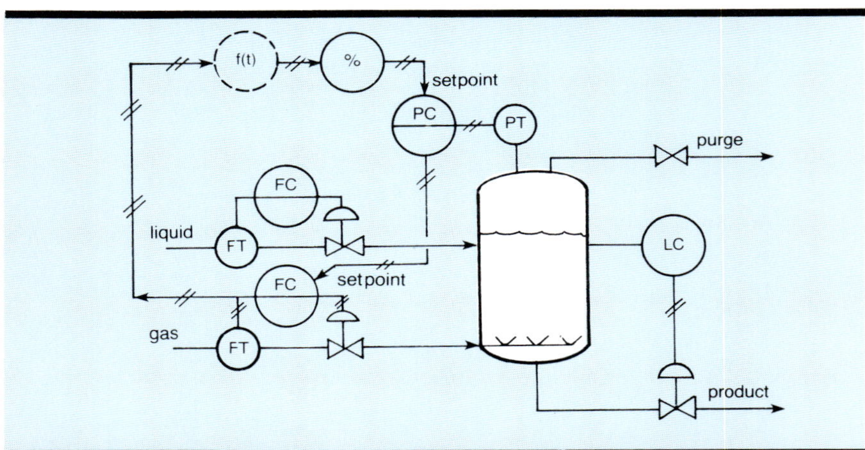

Fig. 3-4. Because the gas phase is consumed in the reaction, pressure can be used to control endpoint.

There are additional considerations in this type of reaction, however. If the gas feed contains inert components they will tend to accumulate in the gas space thus reducing the partial pressure of the reactant. This will reduce the reaction rate so that more of the liquid reagent will leave the reactor unconverted, with its concentration gradually increasing. Figure 3-4 shows a small purge stream being removed from the gas space to withdraw inerts. They will then no longer accumulate but reach an equilibrium concentration which is a function of purge flow, and gas feed rate and composition. Purge flow must be kept low to minimize reactant loss.

This liquid-phase reaction could benefit from residence-time control, as described in Fig. 3-2. Otherwise, the reduction in residence time experienced by a constant volume at increased production will cause the concentration of unreacted liquid feed to rise. Alternately, partial pressure of the gaseous reactant could be increased in proportion to gas flow through a ratio station. This forms a positive-feedback loop, however, which must be stabilized by higher gain through the pressure controller. (A lag may also be required if pressure response is slow.) Raising pressure with flow also raises purge rate, tending to regulate gas-phase composition.

If the product of the above reaction were a gas, then the pressure controller would have to be used to manipulate product flow. Gas reagent would then be under flow control, and the liquid-level controller would maintain endpoint by adding the required flow of liquid reagent. In this case, a liquid purge would be necessary to remove inerts, such as lubricating oils. Residence time can no longer be regulated by a proportional-level controller because its action would be reversed. Instead, a proportional-plus-integral controller must be used, with level set in ratio to flow rate.

When there are products leaving in both phases, the level and pressure controller are both occupied by controlling product inventories and neither can be used to maintain endpoint.

3-6. Recycle Streams

Many reactions must be moderated by the presence of a solvent, a product, or by the excess of one reagent. It then becomes necessary to return the moderator to the reactor after the other components are removed from it. This usually works quite well *if* all other components are, in fact, removed. However, any that remain will tend to accumulate as the moderator is recycled again and again.

In the case of reactant recycle, production rate is controlled by the flow of the feed that is totally converted into product. The excess reactant is simply set in ratio to it, as shown in Fig. 3-5. After separation from the product, the excess is returned to storage from which it is withdrawn for another cycle. Liquid level in the storage tank will tend to fall as reactant is consumed, so fresh reactant is ordinarily supplied under level control. Storage capacity here is important and must relate to the size of the stream. The vessel must be able to contain all the reactant in the plant during a shutdown condition. Constant-inventory control here is both unnecessary and undesirable. In fact, constant-inventory control of the entire stream is more important, but this must include the variable inventory of reactor and separations unit. If the reactor is under residence-time control, its inventory will be low at low production rates. To compensate, then, the recycle-storage-tank level should be high at low production rates. A proportional-level controller admitting fresh feed to storage can accomplish this end.

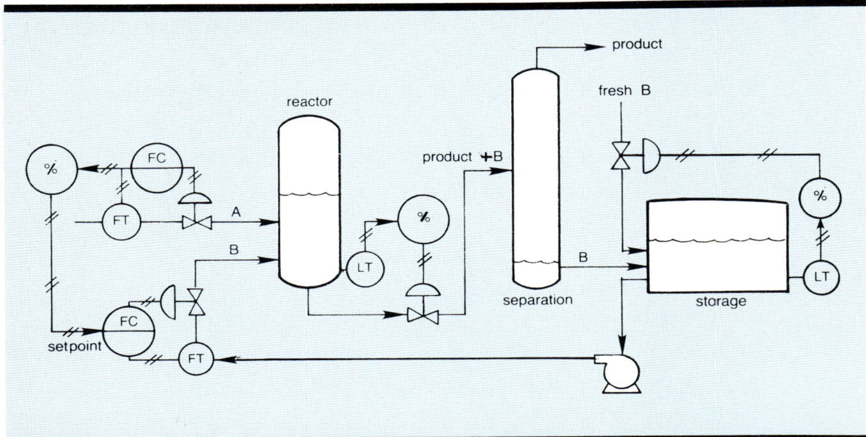

Fig. 3-5. Inventory of excess reactant B will shift from the reactor to the storage tank when production rate is reduced.

Where an inert moderator, such as a solvent, is recycled, total inventory is inherently constant, except for losses which are usually minor. (This is the same situation that exists with any circulating fluid that is not consumed, such as refrigerant, brine, etc.). Then there is no need for inventory control except when the system is initially charged or losses are made up from time to time. (Again, it is a mistake to overcharge the system, or the storage tank may overflow during shutdown.)

Recycle of unconverted reagents in the inert carrier will eliminate the self-regulating character of the material balance. Consider the acid-base reaction without recycle: A deviation of the reagent feed ratio from equivalence will produce a proportional excess of unreacted feed in the effluent, but there will be a steady state. If that excess is recycled with a solvent, however, it will accumulate in concentration, and no steady state will be reached. Endpoint control then becomes essential if a steady state is to be maintained

A recycle system that could be nonself-regulating is the scrubbing of flue gas for sulfur-dioxide removal as shown in Fig. 3-6. Alkaline solution is added from storage under pH control to neutralize the sulfur dioxide absorbed from the flue gas. The spent solution is then regenerated, and waste products are removed before returning the solution to storage.

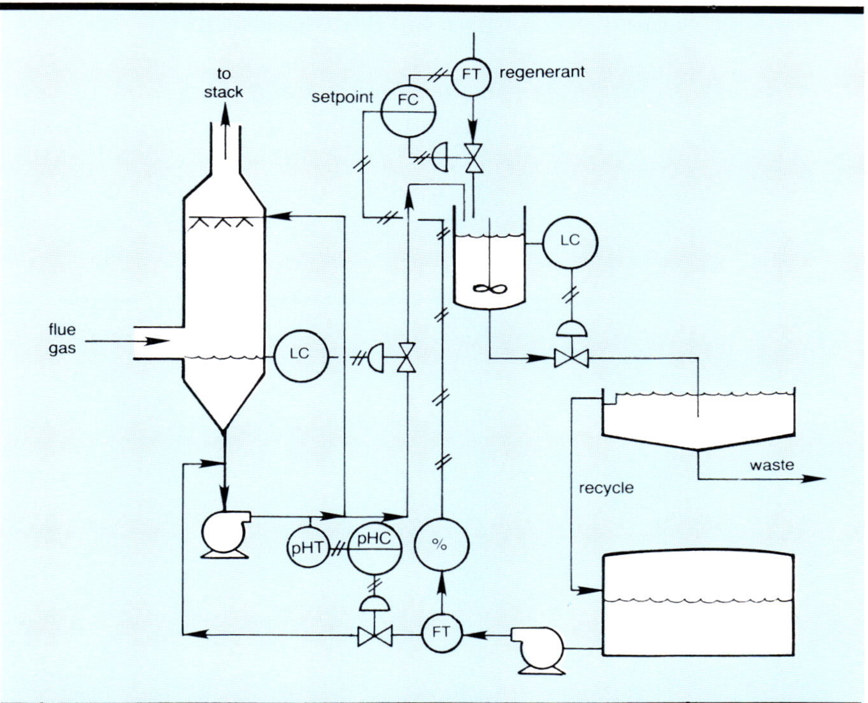

Fig. 3-6. Setting regenerant flow in ratio to demand established by the pH controller regulates solution concentration.

Without end-point control of the regeneration reaction, solution concentration will either rise or fall, depending on the balance between sulfur-dioxide removal and regenerant flow. Figure 3-6 shows the addition of a ratio loop from the flow of solution under pH control to the flow of regenerant. Then variations in solution flow manipulated by the pH controller to meet the demand of the flue gas bring about proportional changes in regenerant flow. Should the *ratio* be increased it will raise the strength of the solution, which will ultimately decrease solution flow when a new steady state is reached.

3-7. Bidirectional Inventory Controls

Unit 2 addressed the location of the production-rate controller and the accommodation of constraints. Transfer of control to other limited manipulated variables was implemented by valve-position controllers. In this section, consideration is given to transfer of production-rate control serially from one process stage to another, with storage vessels between them. Inventory controls for the storage vessels must then be capable of manipulating either inflow or outflow, depending on whether the rate-controlling constraint is upstream or downstream.

Figure 3-7 describes a multistage process whose production rate can be set at either end or constrained at any intermediate operation. If product flow controller FC-3 is limiting, level will rise in tank 3, causing its high-level controller LC-3H to throttle its inflow. This action will raise level in tank 2 to cause LC-2H to act similarly, and LC-1H to manipulate feed rate.

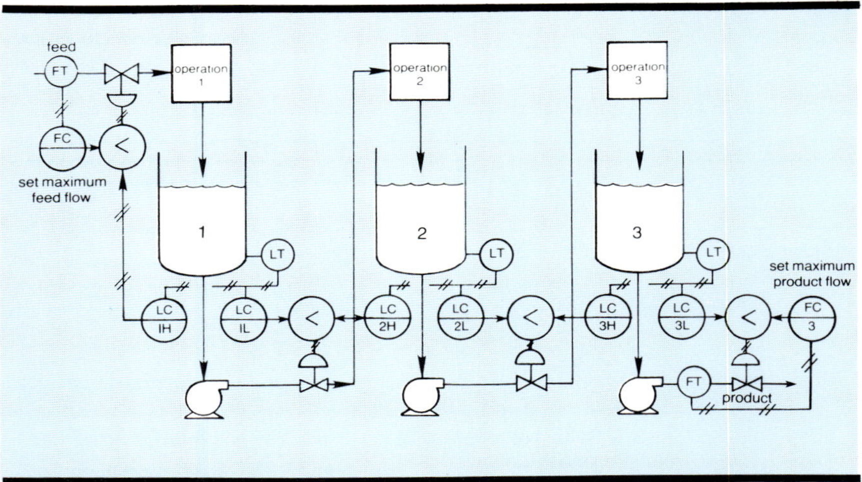

Fig. 3-7. Production rate can be set at either end of the process or constrained at any intermediate point without loss of inventory control.

Should the operator determine that feed rate is too high, he may reduce the setpoint of FC-1 below its measurement, causing FC-1 to take over feed manipulation. The subsequent reduction of inflow to tank 1 below outflow will cause its level to fall. Ultimately, its low-level controller LC-1L will react by taking control of outflow. This action will cause tank-2 level to fall, repeating the same scenario. Eventually a new steady state will be reached at the lower production rate and with lower levels in all tanks.

The system also accommodates constraints at intermediate points. Suppose a filter in operation 2 began to clog, reducing flow into tank 2. Its falling level would cause LC-2L and eventually LC-3L to manipulate downstream flows. Meanwhile, the level in tank 1 would rise, causing LC-1H to reduce the feed to match the rate of outflow.

If all level controllers have integral action, levels will come to rest at either their high or low setpoints, depending on the location of the flow limit. When the limit shifts its location, levels between the two locations will ramp from one setpoint to the other.

If all level controllers have proportional action alone, levels will reach a steady state that is proportional to the plant throughput. Another alternative would have proportional high-level controllers and proportional-plus-integral control of low level. Then levels will be proportional to throughput for vessels upstream of the flow limit and at the low-level setpoint for vessels downstream. With any of these choices, the tank capacities are used for buffering between operations, delaying the transmission of upsets in either direction. Momentary upsets in one operation might not interfere with adjacent operations at all.

Exercises

3-1. *Why is self-regulation a desirable property for a process to have?*

3-2. *Is the level of water in a boiler steam drum self-regulating? Explain.*

3-3. *What would make boiler steam pressure self-regulating? What would make it non self-regulating?*

3-4. *Add a surge tank to the process in Fig. 2-2 and arrange its controls so that the base level controller in the column does not have to manipulate its feed rate.*

3-5. *In what way does the system of Fig. 3-7 provide nonlinear level control?*

3-6. *Devise a bidirectional pressure-control system for a boiler supplying steam to a turbine.*

References

[1]Shinskey, F.G. *Process-Control Systems*. 2nd Ed. New York: McGraw-Hill Book Co., 1979, p. 20.
[2]*Ibid.*, pp. 116-118.
[3]*Ibid.*, pp. 253-257.
[4]*Ibid.*, pp. 134, 135.

Unit 4:
Environmental Variables

UNIT 4

Environmental Variables

This unit describes the control of the environmental conditions in which process operations take place and how the environmental variables may be adjusted to improve process performance.

Learning Objectives — When you have completed this unit, you should:

A. **Recognize those conditions which affect process performance.**

B. **Be able to apply environmental controls to protect a process from disturbances in source and sink.**

C. **Understand how to adjust environmental variables to maximize process performance.**

4-1. Relationship with Inventory Variables

Many of the variables which identified inventory in Unit 3 are also environmental variables, but not all of them. Environmental variables are distinguished by their effects on process performance. The temperature in a chemical reactor may represent an inventory of heat, but it also profoundly affects the rate of reaction. By contrast, the level of liquid in a storage tank may *only* represent inventory in that it has no effect on process performance.

The presence of self-regulation often identifies the environmental variables in that, as they change, they cause proportional changes in inflow and/or outflow. But environmental variables also affect product quality and the efficiencies of process equipment and machinery as well. In many instances, control over the process environment is sufficient to regulate product quality. In most cases, however, quality control requires that a particular relationship between the environmental variables and production rate be maintained. This is exemplified in a dryer, where air temperature is the environmental variable having the principal influence on product moisture. However, changes in production rate, feed moisture, and ambient humidity all require air temperature to change in a prescribed manner if product quality is to be controlled.

There is generally more than a single environmental variable which affects product quality—temperature, pressure, and composition all have their own influences. It then becomes possible to program these variables with respect to each other in such a way as to maintain product quality while improving process performance. For a reactor, this might amount to maximizing yield or catalyst life. For a separations unit, it would result in minimizing energy requirements. Each individual process needs to be examined to determine which of the variables that can be measured and controlled has the most pronounced effect on product quality, and how they can be coordinated to maximize efficiency.

4-2. Equilibrium vs. Nonequilibrium Conditions

Some processes require environmental controls over equilibrium conditions; in general, those processes have insignificant rates. Such is the case with material that is stored for extended times, as food in a freezer or paper in the process of being printed for instrument charts. Their environment largely determines their quality, but there is little transfer of material or energy between the product and its environment.

The same might seem to be the case for a room environment where people are at work, but there is a difference. People always radiate heat in proportion to their activity. Therefore, an environment where heavy work is being done needs to be cooler than a space where people are at rest. Even here, the quality of the product, which is comfort, is affected by how people dress as well as the environment. Then, there is an opportunity for optimization in the form of energy savings by adjusting the combination of clothing and temperature while keeping comfort constant.

Most processes operate in a nonequilibrium because production must take place at a definite rate. Environmental variables then become the rate-controlling mechanisms relating product quality and quantity. Temperature, pressure, and composition determine the rate of a chemical reaction, the rate of drying of a solid, the rate of heat and mass transfer in most processes. These are important considerations for the control engineer, for his system must be capable of controlling a process not only at design conditions, but also under all possible sets of conditions which may be imposed upon it.

Production rate is the most important concern because it can range from zero to full load, and can change virtually instantaneously. Feed composition tends to be limited to a narrower range and can change only as rapidly as source capacity will allow. Ambient conditions represent a significant set of disturbing variables to those units which reject heat into the atmosphere, e.g., refrigeration units, or draw feed from it, e.g., air compressors. Ambient variations can cover a full range of load for some processes like refrigeration and even span two distinct load ranges in the case of heating and air-conditioning. Separate systems required for the two distinct operating modes need careful coordination to minimize energy usage.

4-3. Reaction Conditions

The rate of a chemical reaction varies exponentially with temperature. This is true of competing reactions as well as the desired ones and of reactions involving degradation of the product as well as its formation. As a result, many performance criteria such as product quality, yield, and catalyst life depend on precise and responsive control of reactor temperature.

Endothermic reactors are highly self-regulating and therefore easy to control by the application of heat. Some exothermic reactors may be self-regulating, but most have negative self-regulation—presenting the possibility of a runaway reaction. Self-regulation depends on the heat of reaction, reactant concentrations, and the heat-removel mechanism. (An examination of reactor stability is beyond the scope of this work—for further information the reader is directed to Ref. 1.)

Temperature is a measure of the energy stored in the reactor and can be controlled either by heat input or heat removal. Heat input is directly proportional to reaction rate, which, in turn, is proportional to reactant concentration. If reactant concentration is very low or if the residence time is quite short, owing to a very fast reaction-rate coefficient, temperature can be controlled quite effectively by manipulating feed rate. This is not the normal loop arrangement, however, in that production rate would then be set by the rate of heat removal.

To avoid this problem, in some reactors temperature is controlled by introducing a diluent, which reduces reactant concentration and cools through sensible heat at the same time. Figure 4-1 describes a hydrocracker which is controlled by diluting the reaction mass with excess hydrogen. Inlet temperature is regulated at the feed heater—this is necessary to initiate the reaction. In a multiple-zone reactor, it is possible to optimize the temperature profile for the best combination of yield, production rate, and catalyst life.

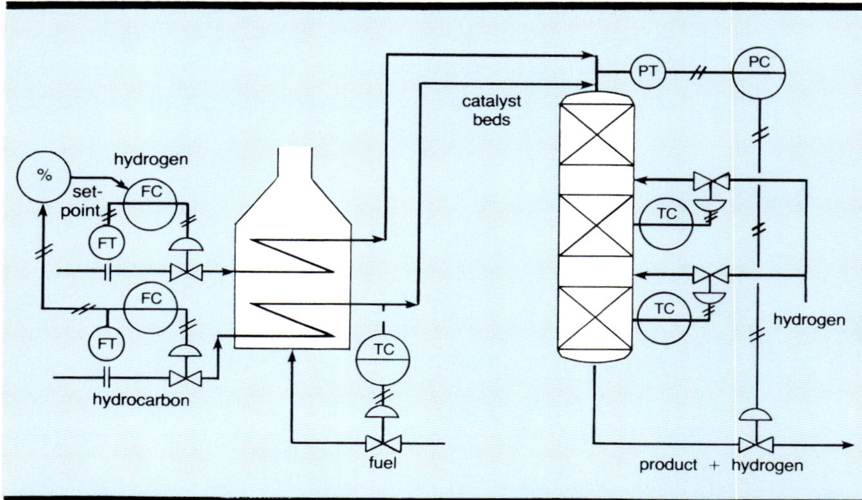

Fig. 4-1. Temperature is controlled at each of several zones in the hydrocracking reactor by dilution with cool hydrogen.

In many reactions, particularly those that function in the liquid phase, the residence time of the rate-controlling ingredient may be minutes rather than seconds. Then the response of reactor temperature to adjustments in feed rate is retarded by the intermediate step of altering reactant concentration. When the concentration time constant approaches or exceeds the thermal time constant, temperature control over feed rate become too slow for stable performance. This is especially true for reactors without backmixing. Then manipulation of heat removal is mandatory.

Figure 4-2 shows a jacketed reactor whose temperature is controlled by manipulating coolant exit temperature in cascade. For maximum stability, coolant must be continuously recirculated, with cold water added for temperature control. Hot water or steam is necessary for startup. The reactor temperature controller needs all three modes, while proportional or proportional-plus -derivative are preferred for the jacket controller.

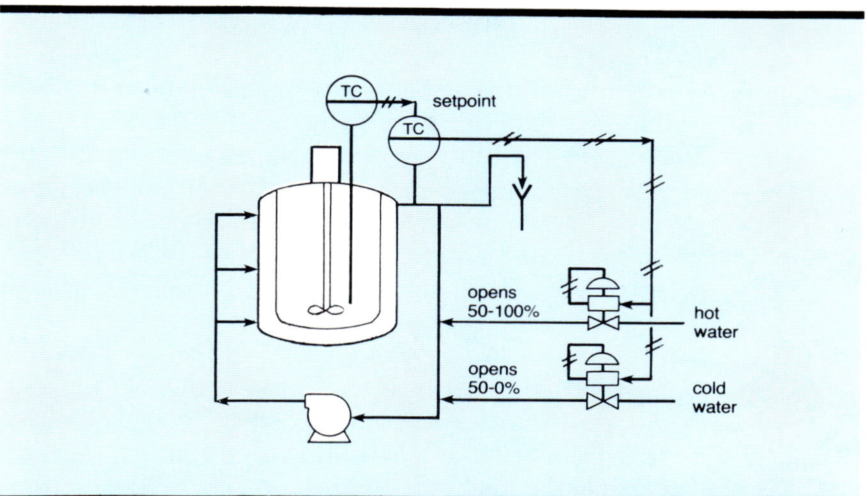

Fig. 4-2. Stirred-tank reactors need cascade control of recirculated cooling water for stable temperature regulation.

An even more stable heat-removal method is to boil water in the jacket as shown in Fig. 4-3. Then temperature is controlled by setting the jacket pressure, which determines the boiling temperature.

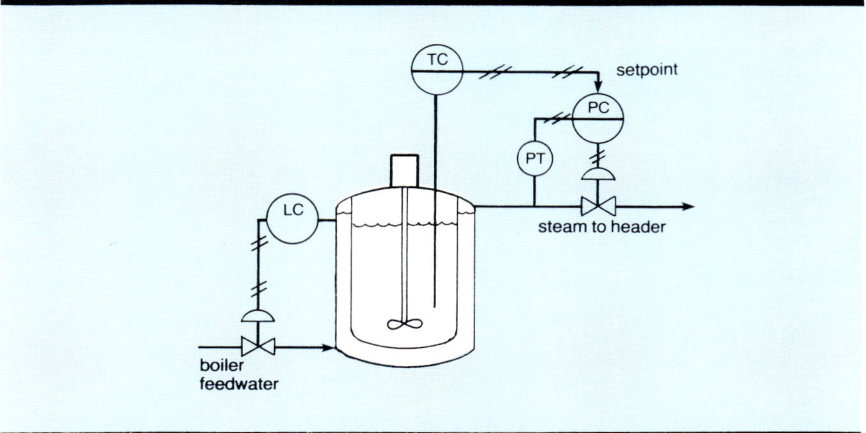

Fig. 4-3. A boiling coolant stabilizes an exothermic reactor by providing an isothermal heat sink.

Most reactions involving gases are sensitive to pressure as well as temperature. If the volume of the products is lower than the volume of the reactants, then increasing pressure will force the equilibrium in the direction of increasing production. These reactions are usually conducted at high pressure—ammonia synthesis from nitrogen and hydrogen is an example. If the volume of the products exceeds the volume of the feeds, then lower pressure will promote the reaction. Again, there are usually competing or serial reactions to be considered, such that there exists an optimum pressure as well as an optimum temperature. If the reaction is conducted in a boiling liquid then pressure and temperature may not be independent, so that only one need be controlled. In this case, pressure is usually chosen for its faster response.

Reaction equilibrium is actually governed by partial rather than total pressure. Therefore, the accumulation of inert gases in a mixture at constant total pressure brings a reduction in the partial pressure of the reactive gases. If it is necessary to control the conditions which determine reaction equilibrium, then partial pressure must be controlled. This can be implemented by controlling total pressure and concentration of the reactive gases, if a means is available for concentration control. If it is not, then partial pressure should be calculated by multiplying total pressure by mole fraction, and then controlled in the same way as total pressure. Then variation in the composition of the gas mixture will cause total pressure to rise or fall proportionately.

Control of concentrations within the reactor was discussed under the heading of Endpoint Control in Sec. 3-5.

4-4. Regulation of Sources

Material and energy are applied to various process units from sources which require some degree of regulation. Pumps and compressors usually are under pressure control, and heat-exchangers and furnaces are under temperature control. The regulation of the source of supply is crucial particularly when there is more than one user. If the source were rigidly regulated, then variations in demand by one user would not affect others at all. If it were unregulated, an increase in demand by one user would result in a proportional decrease in supply to all others to a degree based on the self-regulation of the source.

Consider what would happen if the source had no regulation of any kind, such as a positive-displacement pump supplying two users. A change in demand by one user would impose an equal and opposite change on the other. A centrifugal pump, on the other hand, has self-regulation in the relation of head to flow. An increase in flow to one user will decrease flow to the other through a reduction in head; but the lower head is the result of an increase in total flow, so that the second flow is reduced less than the first is increased. If the interaction is to minimized, pressure control must be provided. Even then, interaction may not be eliminated altogether, in that time is required for the pressure controller to function, during which a transient disturbance may be observed. Feedforward control can be helpful here, especially in furnaces, where delays are longer.

Pressure and temperature represent the inventory of mass and/or energy at the point of delivery between the supply and demand. But they are also environmental variables in that they affect the performance of both upstream and downstream processes. The temperature of the reactor feed in Fig. 4-1 determines the reaction rate, but it also impacts on the efficiency of the heater. Higher temperatures and longer residence times in the heater increase heat losses and degrade both the feedstock and the heater tubing.

The same is true of pressure controls on pumps and compressors. Higher discharge pressure and lower suction pressure may affect adjacent processes, but they also increase the power required per unit of throughput and the wear on mechanical components.

The difficulty encountered in regulating a source of supply depends as much on the quality of the source as it does on the nature of the demands. If fuel to a heater is of a constant quality, then the controls only need react to variations in demand. But if source quality such as fuel is variable, this too will have its effect on the controlled variable. If quality variations are slow to develop, feedback control over the environmental variable may be adequate. If not, the source may have to be compensated. In the case of fuel quality, compensation must be made for both heating value and density—in a function known as the Wobbe Index (2). The speed with which compensation is applied must be faster than the response of the process being upset—if not, compensation will be too late to be effective. Figure 4-4 shows a Wobbe Index analyzer being used to calculate the heat flow being delivered by a fuel valve. Note that the temperature controller sets heat flow rather than fuel flow.

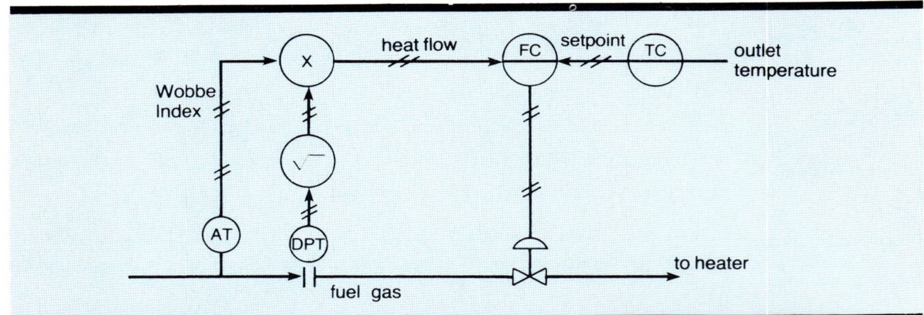

Fig. 4-4. Compensation for variations in source quality may be applied to the flow measurement.

4-5. Regulation of Sinks

The environment is our universal sink for both material and energy. It cannot be regulated, for its temperature, pressure, and composition are quite beyond our control. Yet there are various avenues through which material and energy are rejected into the environment, and these avenues frequently need regulation. Liquid wastes need to be treated by neutralization, oxidation, etc., and gaseous wastes need to be incinerated, scrubbed, etc. prior to discharge.

But there are far more point discharges of energy than material, and different avenues of heat rejection, such as river water, air, cooling towers, and refrigeration units. Isolation between the process and the environment is provided by a heat-transfer surface which presents a temperature difference proportional to heat flow. Furthermore, the temperature of the environment is variable. If process temperature or pressure is to be held constant, some means of control is necessary to adjust for the effects of variable heat flow and temperature of the coolant.

Dry atmospheric cooling is most variable because air temperature is most changeable, particularly in temperate climates. Wet atmospheric cooling is less variable because it is determined by ambient wet-bulb temperature which changes less than dry-bulb temperature. Surface water and groundwater are still less variable owing to the heat capacity of the earth. A refrigeration unit is a heat pump between the process and the environment, which, because of additional heat-transfer stages, becomes more variable than the environmental sink into which it discharges.

The difficulties encountered in regulating heat sinks is typified by the distillation column condenser shown in Fig. 4-5. Condensing temperature and therefore saturation pressure are determined by air temperature and the gradient across the heat-transfer surface. As air temperature and heat load change, saturation pressure will rise or fall, requiring an adjustment to the heat-transfer rate. The bypass valve accomplishes this by changing the amount of liquid held up in the condenser, thereby affecting the surface area available for condensing duty. Speed of response is relatively slow because liquid level within the condenser must change to affect the rate of condensation. Yet, because of its linearity and reasonably wide range, this is one of the preferred means of condenser control. Throttling coolant flow is nonlinear and promotes fouling; injecting noncondensible gas augments product losses.

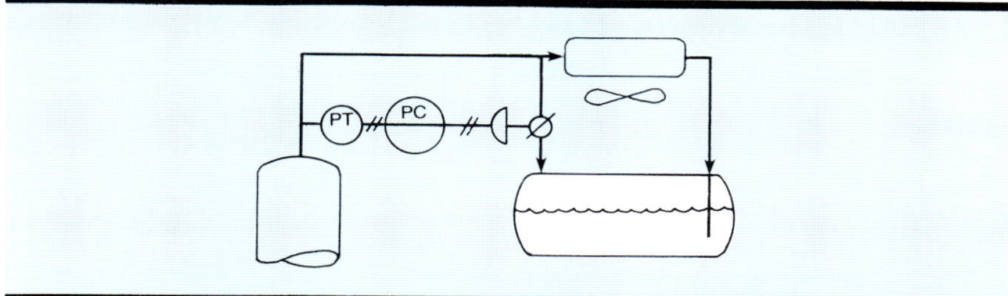

Fig. 4-5. The hot-vapor bypass valve reduces the rate of condensation by holding liquid in the condenser.

An important fact to realize is that the controls applied to regulate the heat sink need to be carefully matched to the characteristics of the sink. There are so many varieties, even of distillation column condensers, that they cannot even be mentioned here. For further information, the reader is directed to Ref. 3.

4-6. Optimizing Environmental Variables

Because environmental variables affect process performance, it is possible to program them to optimize performance. However, accomplishing process optimization is not easy—it requires an intimate knowledge of relationships between variables and the characteristics of plant equipment that its designers might not even have.

Optimization may take many forms. In some cases it involves simply setting the setpoint of a controller at some maximum or minimum limit, knowing that equipment constraints prevent operation beyond that point. In batch processes, optimization may require a particular program of a controlled variable, such as temperature as a function of time. Other variables may condition that function, however, such as initial temperature and concentration. Continuous processes frequently require programming of a controlled variable against an uncontrolled variable, such as temperature against concentration.

Sources and sinks generally are optimized at constraints which vary with flow rate and environmental conditions. Consider the air compressor supplying several users in Fig. 4-6. Discharge pressure should be minimized to save compressor power, but it must be high enough to satisfy the most demanding user. Demand is measured as the position of the user control valve. The most-open valve is selected for valve-position control at a setpoint of 90% or so. This allows a 10% control margin to accept increasing load changes. The valve-position controller must have integral-only action—proportional action would pass upsets from the selected user directly to the pressure controller and thereby upset other users. Closure of the valve-position loop depends on the action of the pressure controller and the response of the user flow controller in reacting to changes in supply pressure. Hence the VPC must be slower than the combination of those two loops, but will minimize energy consumption in the steady state.

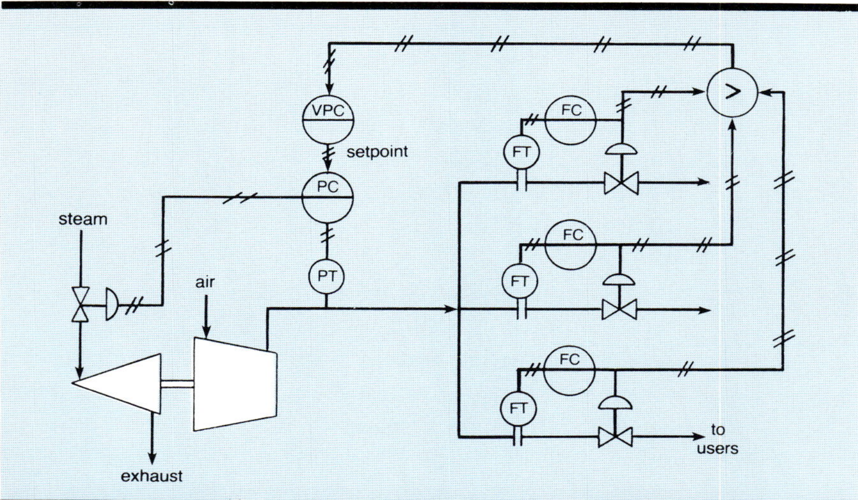

Fig. 4-6. The valve-position controller lowers pressure setpoint until the most-open valve is nearly wide open.

Figure 4-7 illustrates optimization of a heat sink for a distillation column. Most mixtures require less energy to separate by distillation as pressure and temperature are reduced (3). The lowest pressure at which the column can be controlled is that at which the bypass valve is nearly closed. The valve-position controller lowers the pressure setpoint until the valve is at its lowest controllable opening, e.g., about 10%. Its action must be very slow, with an integral time of about an hour, to allow the heat capacity of the column contents time to change to the new boiling point. Proportional action cannot be used.

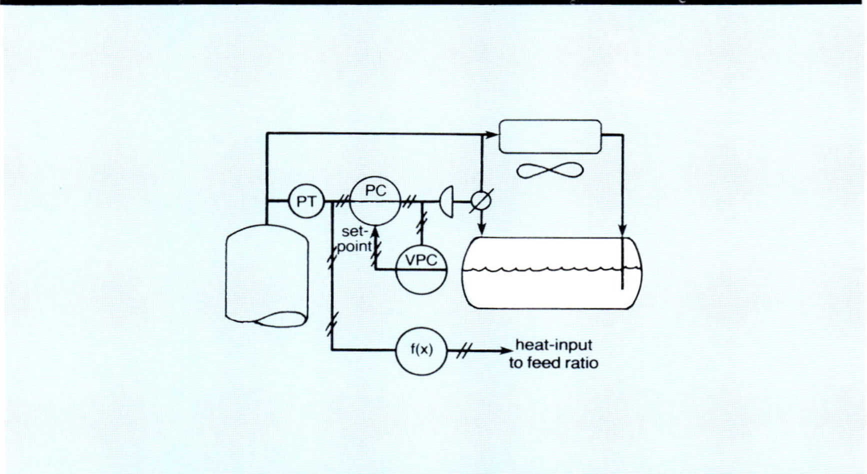

Fig. 4-7. The valve-position controller sets the pressure setpoint to its lowest controllable value.

Varying column pressure saves no energy by itself—it only will cause the purity of the products to improve. If energy is to be saved, heat input needs to be readjusted to maintain constant purity. This can be accomplished by product-quality controllers, but tends to be quite slow due to delays in product sampling and analysis. Response time is shortened by feeding the pressure signal forward to the heat-input controller through a linear function generator which is matched to the characteristics of the column. For further discussion on this application, see Ref. 3.

Exercises

4-1. *What makes a centrifugal pump self-regulating?*

4-2. *Why is a heat source necessary for an exothermic reactor?*

4-3. *Devise a system to control partial pressure of hydrogen at the inlet of the hydrocracker in Fig. 4-1.*

4-4. *Why is it necessary to control pressure in a distillation column?*

4-5. *When should an environmental variable be compensated for rather than controlled?*

4-6. *The temperature of hot oil used to heat a column reboiler is variable. How can a constant heat flow be obtained?*

4-7. *What variable(s) in a refrigeration system may be optimized to save energy?*

References

[1]Shinskey, F.G. *Process-Control Systems*, 2nd Ed. New York: McGraw-Hill Book Co., 1979, pp. 253-257.
[2]Shinskey, F.G. *Energy Conservation through Control*. New York: Academic Press, 1978, pp. 42, 43.
[3]Shinskey, F.G. *Distillation Control: for Productivity and Energy Conservation*. New York: McGraw-Hill Book Co., 1977, Ch. 7.

Unit 5:
Product Quality Controls

UNIT 5

Product Quality Controls

This unit presents the importance of responsive and precise control over product quality, along with the difficulties encountered in obtaining it, followed by suggestions for overcoming those difficulties.

Learning Objectives — When you have completed this unit, you should:

A. **Appreciate the economic incentives for quality control.**

B. **Understand the nonlinear and dynamic elements in the quality-control loop.**

C. **Be able to devise systems capable of achieving control in the presence of these unfavorable elements.**

5-1. Quality and Cost

There is a very strong relationship between product quality and cost—a relationship which is hyperbolic or logarithmic in nature, as shown in Fig. 5-1. To increase product purity beyond what is absolutely necessary can be very costly. The cost penalty may accrue in a variety of ways. Increased purity often may be obtained by using more energy per unit of product made. If energy flow is already at a maximum, then increased purity requires a concurrent reduction in product flow, thereby reducing plant capacity. If energy flow per unit product is not to be increased, then quality can be improved only by reducing product recovery, in which case losses are increased.

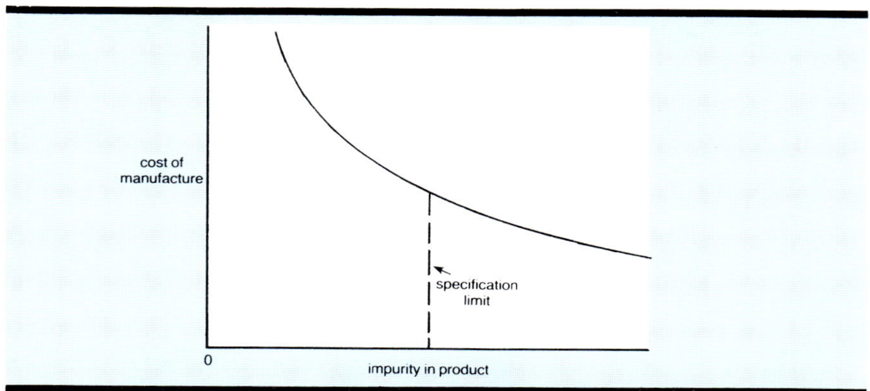

Fig. 5-1. The cost of manufacturing a product increases drastically as absolute purity is approached.

All of the above are undesirable, yet the importance of meeting specifications cannot be ignored. If a product fails to meet specifications, other costs accrue in that the product can no longer be sold for the intended purposes. It possibly may be sold for less-demanding purposes, but at a lower price. If no market exists for the lower-grade product, it may have to be rerun, blended with above-grade product, or destroyed. Disposal costs are now quite high, so that route is becoming less attractive. Reprocessing increases operating costs substantially, and is only used when quality is too low for blending.

Blending with above-grade product can be accomplished in a number of ways, but all impose the same penalty. Because the cost-vs.-impurity curve is nonlinear, manufacturing an amount of above-grade product sufficient to offset the poor quality of the below-grade product costs more than if the specification product were made directly. By the same reasoning, cycling product quality on equal sides of the specification is also more costly than holding specification exactly, in proportion to the amplitude of the cycle.

In practice, the quality of most products varies somewhat randomly as a function of disturbances in feed and ambient conditions, or may cycle as shown in Fig. 5-2. If quality control is effective, the controller will minimize these excursions, allowing the setpoint to be moved closer to specifications. But due to delays in product analysis and the high sensitivity of many processes to disturbances, quality variation may be extensive, requiring a large operating margin between the setpoint and the specification. Improved quality control then represents a substantial opportunity for cost reduction.

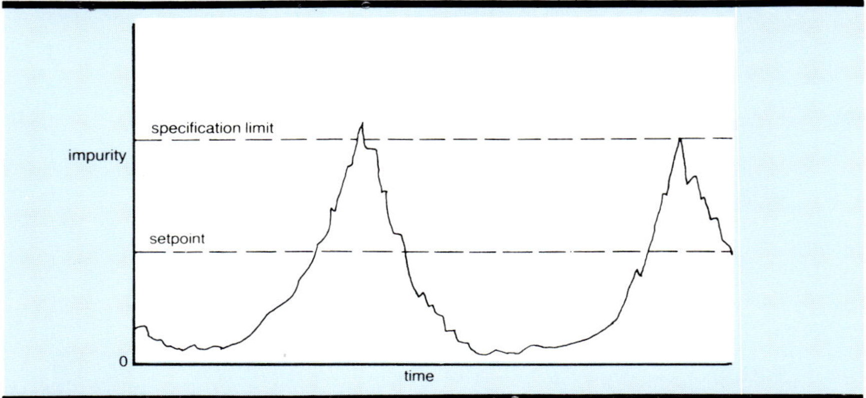

Fig. 5-2. To keep impurity levels from exceeding specifications, the setpoint is set below specifications in proportion to the amplitude of observed variations.

5-2. Determining Product Quality

Precise determinations of product quality have been made in the analytical laboratory, and in some cases, this is the only available determination. In recent years, however, many analytical instruments have moved into the field, where they have been directly connected to the process being controlled. In most cases, control of a particular component in the product is necessary, requiring that a specific analysis be made. Then some device which can detect that component against the background of many others is required.

The techniques most commonly used for on-line chemical analysis are gas chromatography and photometric absorption. The chromatograph separates a discrete sample of the mixture into its components, detecting the relative quantity of each by comparison of a physical property, such as thermal conductivity, against the background of a carrier gas—usually helium. For light hydrocarbons, chromatographic analyses are usually quick, sensitive, and reliable. Separating higher-boiling and close-boiling mixtures becomes more difficult, time consuming, and expensive.

Photometric spectral analyzers detect the presence of a component by its electromagnetic absorption or transmission in a particular portion of the infrared, visible, or ultraviolet spectrum. Absorption is a function of bonds between atoms, and so these instruments are insensitive to elemental components. On-line spectrometers operate continuously rather than discretely, but are usually large and delicate, so that they need to be protected from the environment in a shelter of some kind.

Ion-selective electrodes are devices which develop a potential that varies with the logarithm of the activity of one or more ionic species in contact with them. Being logarithmic, their sensitivity increases with falling concentration, making them very valuable for trace analysis in pollution control. The pH electrode is their foremost representative, but electrodes sensitive to many other ionic species now exist. A gas-phase version sensitive to oxygen is in common use for flue-gas analysis.

Other ionic measurements include reduction-oxidation potential, useful for controlling endpoint in certain liquid-phase reactions, and electrolytic conductivity, which is sensitive to all ions present.

Physical-property measurements such as density and thermal conductivity can be used to infer the concentration of a component in a binary mixture or in more complex mixtures where other components remain in fixed ratios. Sometimes a physical property itself needs to be controlled, based on the service that the product will see. Thus, vapor pressure of gasoline is often controlled, or flash point of jet fuel. Boiling point is commonly controlled in distillation columns, as an inference of composition. Whether these inferential measurements are sufficiently representative of product quality depends on many factors, such as pressure, other impurities, and the sensitivity of the measurement itself. Differential measurements comparing the product to a known reference improve accuracy.

5-3. Dynamic Response

Most processes wherein product quality is controlled consist of a series of stage-wise operations or a longitudinal distribution of product.

Distillation columns have many trays, evaporators usually consist of several serial effects, and most dryers transport product relatively long distances. In order to change product quality, flow rates of product, feed, reflux, or energy need to be adjusted. Within these operations there exist strong interactions between material balances and point compositions, often requiring several iterations to reach a steady state following a disturbance. The associated delays between flow changes and resulting composition changes pose difficulties for the product-quality controller. The controller cannot impose changes more rapidly than the process can respond, and, in fact, must wait for response before proceeding. These delays make the process overly susceptible to disturbances which occur faster than the quality controller can react to them. The sensitivity of product quality to virtually all the variables in the plant further compounds the issue.

If a sample must be withdrawn from the process for analysis, its transport time to the analyzer adds directly to the dead time of the process itself. The chromatograph has two additional undesirable dynamic features. The sample is delayed in moving through the analyzer to the detector, during which time component separation takes place. Furthermore, the analyzer samples discretely rather than continuously, introducing another delay equal to half the sampling interval (1).

Photometric analyzers have no internal dead time, and only a short time constant—still they require transportation of the sample from the process to the analyzer. Care should be exercised in designing the sampling system so that transportation time is minimized. Vapor rather than liquid samples should be used whenever possible, because they can travel with greater velocity. Also, the analyzer should be located as close to the process as possible.

The most responsive systems are those that do not require a sample to be withdrawn. Immersion and in-line pH measurements give much more satisfactory control than those located in sample chambers. This is one important reason why temperature is used so much for distillation control, although it is much less sensitive and accurate than an analyzer. A temperature-control loop on a distillation column may cycle at a natural period of about 20 minutes, whereas the period of a chromatographic analysis control loop on the same column is more like 60-90 minutes.

5-4. Sampled-Data Control

Some of the dynamic disadvantages associated with the sampling nature of the chromatograph can be overcome by using a sampling controller. In operation, the chromatograph gives analytical results only at specific time intervals. Between these intervals, the product-quality loop is open, although a conventional controller will continue to integrate any existing deviation from setpoint.

Improved performance is attainable by acting on the new analysis with all three control modes for one instant only. The controller is then idle until the next analysis is reported. An analog controller can be converted to sampled-data operation by transferring it from manual to automatic for a short time, during which the new analysis is received. The controller is allowed to remain in automatic only for a few seconds, after which it is returned to manual. Proportional and derivative action take place as the new analysis is received, and integral action is exercised during the time the controller remains in automatic. Integral time should be set approximately the same as the time the controller is in automatic.

For digital control systems, the PID algorithm may be executed only once, as soon as an analysis is reported. It is then idle until the next report. Failure of the analyzer to report will not jeopardize the system in that no control action will be taken. This same procedure can be applied to the control of any off-line analysis. The operator may enter new information and execute the algorithm only once. Variations in the time between entries are not generally detrimental to the functioning of the system.

5-5. Splitting the Control Function

Most product-quality variables are sufficiently difficult to control that the conventional PID function may fail to achieve satisfactory results. The designer must then make use of other functions such as cascade, ratio, and feedforward. This section examines another alternative—splitting the control function into its separate modes that are then applied to separate but related variables.

Consider the possibility of two different *measured* vairables, one of which has a rigid steady-state relationship to product quality but poor dynamic response, the other whose steady-state relationship is tenuous, but which responds faster to manipulation. An example is the control of an exothermic reactor, where temperature may have the necessary steady-state relationship to product quality, but does not respond to cooling as quickly as pressure. However, under constant temperature, pressure may vary with the composition of the feed and the condition of the catalyst. The integral mode must then be applied to the temperature measurement to assure that its setpoint is maintained in the steady state. Derivative can be applied to the pressure measurement in that it contributes only to the unsteady state. Thus pressure can be used to improve stability and react quickly to disturbances, although temperature remains as the controlled variable.

A system applying derivative action to a secondary measurement is shown in Fig. 5-3. Differentiation of the secondary measurement, followed by summation with the primary measurement, can be combined in a single device or algorithm—thus they are shown combined. The derivative amplifier may have both an adjustable gain and time constant, in contrast to the derivative action in most controllers, where the gain is fixed at 10-15. The gain should be set as high as measurement noise will allow, with the time constant set for stability of the secondary control loop. Derivative time in this configuration may be much shorter than integral time, which has to be set to favor the slower response of the primary variable.

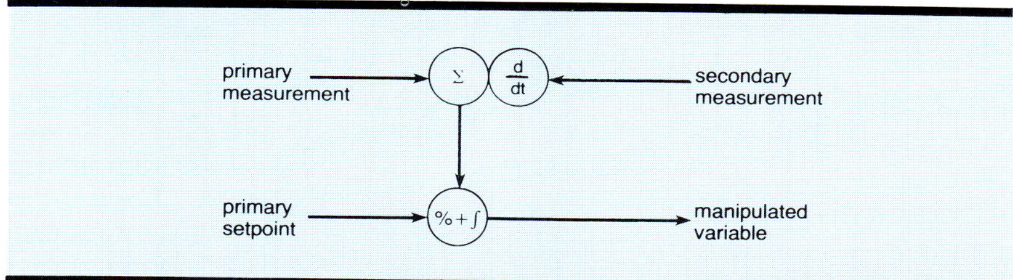

Fig. 5-3. When a secondary measurement responds faster to manipulation than the primary, it may be used for the derivative mode alone.

Another opportunity for splitting the control function appears when there is a secondary *manipulated* variable which produces faster response than the primary but is limited in its range. The secondary manipulated variable also may be more costly to use than the primary, so that its use should be minimized for economic reasons.

Figure 5-4 shows two controllers operating on the same controlled variable and setpoint. Proportional-plus-integral action is applied to the slower, less costly manipulated variable, and proportional -plus-derivative action to the faster, more costly variable. In the steady state, when the controlled deviation is zero, the output of the PD controller always returns to the same value. That value can be positioned by an internal bias in the controller to minimize the steady-state contribution of the secondary manipulated variable. The bias should not be set at zero, however, for then derivative action can be applied only to deviations on one side of the setpoint.

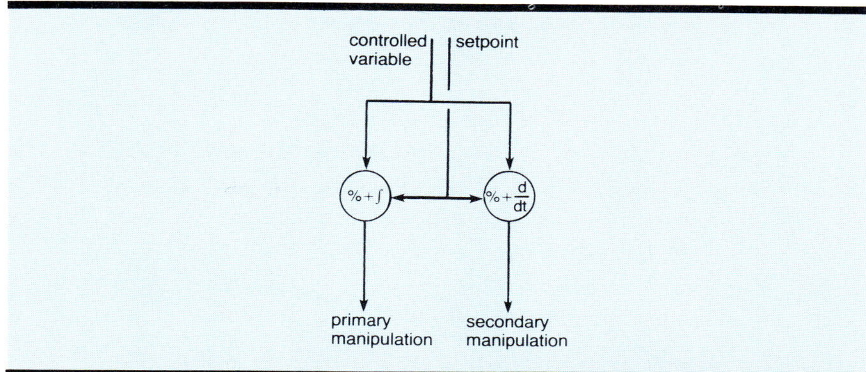

Fig. 5-4. When a secondary manipulated variable produces faster response than the primary, it may be driven by a proportional-plus-derivative controller.

The configuration has been successfully used to control product quality in multiple-effect evaporators (2). The primary manipulated variable is steam flow to the first effect, while the secondary is steam flow to a small trim heater on the last effect. Steam to the trim heater elicits faster response in product quality, but is more costly in that it is used in only the one effect, whereas heat applied to the first effect passes through all. Another variation to the system uses a trim feed stream which bypasses the first effects and enters the last directly. In both cases, the secondary manipulated variable has a very limited range and should return to a minimum controllable position whenever a steady state is reached

5-6. Gain Compensation

Because product quality tends to be so difficult to control, maintaining a constant loop gain is necessary for tight regulation. Yet many processes are nonlinear in one or more dimensions, causing stability to be variable unless some sort of compensation is applied.

In the case of pH or any of the other ionic measurements, the controlled variable changes nonlinearly with composition. Since the manipulated variable brings about composition change, gain compensation needs to be applied at the input of the controller. This takes the form of a nonlinear function applied to the input of the controller, as described under Endpoint Controls in Sec. 3-5.

Many control loops encounter gain variations as a function of flow. Consider a heat exchanger wherein liquid flowing at rate W_p is heated from temperature T_1 to T_2 by condensing steam at rate W_s. A heat balance gives:

$$W_s \Delta H_s = W_p C_p (T_2 - T_1) \qquad (5-1)$$

where ΔH_s is the latent heat of the steam, and C_p is the specific heat of the liquid. Differentiating the controlled variable T_2 with respect to the manipulated variable W_s gives the steady-state process gain:

$$\frac{dT_2}{dW_s} = \frac{\Delta H_s}{W_p C_p} \qquad (5-2)$$

Observe that the gain varies inversely with liquid flow.

If no compensation is applied, reductions in flow will tend to engender cycling due to the increase in loop gain. It is then impossible to adjust the temperature controller for optimum performance at more than one operating point. One solution to the problem is to manipulate steam flow with an equal-percentage valve whose gain increases with steam flow. This is acceptable if the temperature rise of the liquid is reasonably constant, so that steam flow always has the same proportionality to liquid flow. If this is not the case, another approach is needed.

Figure 5-5 shows a scheme where the output of the temperature controller is multiplied by liquid flow. The setting of steam flow proportional to liquid flow gives feedforward action, while the multiplier in the feedback loop introduces gain compensation. Decreasing liquid flow causes the gain of the multiplier, which converts controller output to steam setpoint, to decrease, thus exactly cancelling the gain increase characteristic of the process. This is a natural property of feedforward control in that its formulation is based on the nonlinear process model.

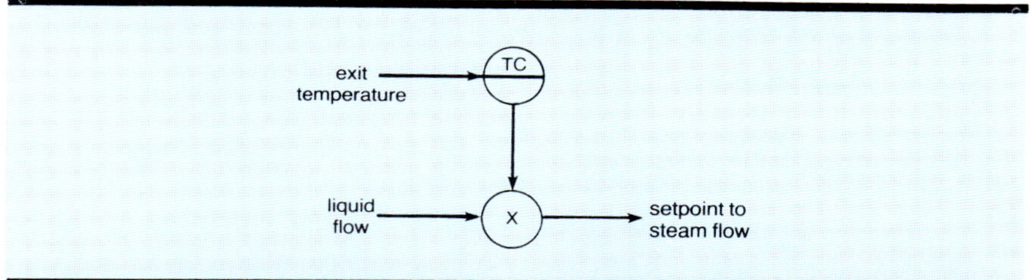

Fig. 5-5. Multiplying controller output by flow incorporates feedforward action with loop-gain compensation.

A very important application of gain compensation is the control of product quality from distillation columns. Reference 3 shows that the relationship between the level of impurity in a product and its manipulated flow varies with the magnitude of the impurity. Thus, as the impurity level approaches zero, loop gain approaches zero, and control action produces very little effect. At the other extreme, high impurity levels cause process sensitivity to increase proportionately, producing a strong corrective response and the characteristic nonlinear cycle shown in Fig. 5-2.

Reducing the gain of the controller will not eliminate the cycle but will only reduce its amplitude and increase the process sensitivity to disturbances. The controller gain needs to be varied in inverse proportion to the level of the impurity. The best way to accomplish this is to adapt the controller's proportional band linearly with the impurity measurement, which is also the controlled variable. The ability to implement the adaptive function is not generally available in analog controllers, although some of the newer, advanced electronic systems have it.

Gain adaptation is common to digital algorithms, since gain or proportional band is simply a stored parameter for insertion in the algorithm and is capable of being calculated as well as entered directly.

5-7. Adaptive Control

The gain compensation described above is classified as adaptive control in that a control-mode setting is automatically adjusted as a function of a changing process characteristic. This type of adaptation is considered to be programmed in that mode adjustment is based on a predetermined process relationship involving a measurable variable. If the gain-changing characteristic of the process is not determinable, or is caused by changes in an unmeasurable variable, programmed adaptation cannot be used.

An example of the latter is the pH control system in which the process titration curve changes shape due to the variable concentration of unmeasured species in the mixture being neutralized. The effects of these concentration changes are observable only in the variable stability of the pH loop. Consequently, if adaptation is to be performed, it must be based on calculations of loop gain or damping.

Adaptive control based on loop stability forms another feedback loop with its own stability considerations. If the gain of the adaptive loop is too high, or its timing too short, it may break into sustained oscillations. Then the primary loop will periodically cycle between undamped and overdamped modes.

Some adaptive systems test the process by injecting periodic disturbances in controller output or setpoint. This practice simplifies the gain calculation but is generally undesirable in its effect on the controlled variable. Without the known disturbance, the loop-stability estimate is much more difficult. Adaptive control has been successfully applied to a nonlinear pH controller by adjustment of the width of the low-gain zone as described in Ref. 4 and in Ref. 1, pp. 158-160. Control was adequate for the waste-treatment process but is not recommended for general product-quality applications.

Exercises

5-1. The energy Q required to operate a distillation column varies with the logarithm of product impurities x and y:

$$Q = -k \ln (xy)$$

At current production, x and y average 0.01 mol fraction, although specifications are set at 0.02. How much energy can be saved by operating at specifications?

5-2. Both x and y cycle between 0.003 and 0.017. Assuming the cycle is a square wave, how much energy could be saved by operating without a cycle, at specifications?

5-3. Since the cycle is not likely to be a square wave, is this estimate liberal or conservative? Why?

5-4. A chromatograph is to be used to analyze the distillate from a column. Where should the sample be taken? Why?

5-5. Devise a control system using column temperature as well as the chromatograph to control distillate composition.

5-6. What advantages does programmed adaptive control have over feedback adaptation?

References

[1]Shinskey, F.G. *Process-Control Systems*, 2nd Ed. New York: McGraw-Hill Book Co., 1979, pp. 103, 104.
[2]Shinskey, F.G. *Energy Conservation through Control*. New York: Academic Press, 1978, pp. 201-203.
[3]Shinskey, F.G. *Distillation Control: for Productivity and Energy Conservation*. New York: McGraw-Hill Book Co., 1977, pp. 258-260.
[4]Shinskey, F.G. "Adaptive pH Controller Monitors Nonlinear Process," *Contr. Eng.*, V. 21 (February 1974).

Unit 6: Control of Economic Variables

UNIT 6

Control of Economic Variables

This unit examines the measures of economic performance for industrial processes and suggests how they might be controlled.

Learning Objectives — When you have completed this unit, you should

A. Recognize economic variables.

B. Be able to apply controllers to those that are measurable.

C. Understand how to optimize those that are unmeasurable.

6-1. Loss Functions

The economy of a process can be stated as the ratio of the value of product made to the cost of making it. It also can be stated in inverse terms, as production cost per unit product made. Other units of measurement often are inserted as cost for purposes of consistency, such as rating a power plant in Btu/h of fuel consumed, per kilowatt of electricity generated. Dimensionless or percentage efficiencies also are reported as output-to-input ratios expressed in the same units—boilers are rated in this way.

However, in measuring efficiency or performance, one should look for losses because they are the cause of inefficiency. Accuracy also is improved if efficiency calculations are based on losses, i.e., the parts of the input that are lost, as opposed to the part that is retained or converted. In the case of an 80% efficient boiler, for example, the accuracy of measuring the 20% energy lost will be about four times as great as that of measuring the 80% converted.

Aside from the measurement or calculation aspects, there is the need to control or maximize economic performance. Again, this can be accomplished best by controlling losses. If losses can be reduced, those savings usually can be passed back to the source of feed or energy, ultimately reducing process input for the same output.

6-2. Controlling Measurable Losses

Many of the losses of materials and energy in a plant are measurable. The loss of energy as measured by oxygen in the flue gas from the boiler in Unit 1 was an example. Although the oxygen measurement represented only the composition component of the loss and not the temperature component, it could be controlled and therefore minimized. Typical of many economic variables, it could not be driven to zero without causing other problems, such as smoke formation and explosion hazard. In fact, there is an optimum oxygen level which takes these hazards into account; the optimum value depends on fuel type, boiler characteristics, and boiler load. Because of the load relationship, the setpoint for the oxygen controller needs to be programmed as a function of measured steam flow. The function is developed from a series of tests and programmed into the control circuit in a curve characterizer, as shown in Fig. 6-1.

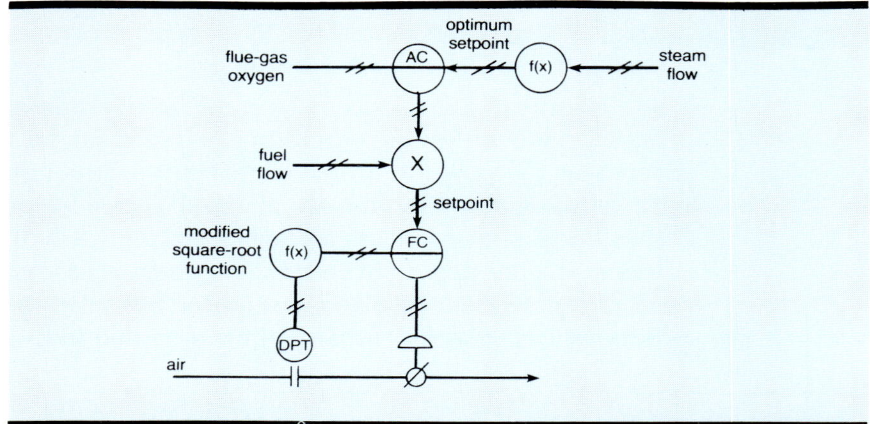

Fig. 6-1. The optimum oxygen content in the flue gas varies with load, requiring appropriate characterization of the setpoint.

It should be noted that this optimum function depends on the condition of the boiler—air leaks and imbalanced burners reduce efficiency at all oxygen levels. These items need to be checked frequently to assure that the programmed setpoint is, in fact, optimum.

Alternately, the concentration of carbon monoxide could be used in place of oxygen as the economic controlled variable, and could manipulate the fuel-air ratio in the same way. Again, there is an optimum value at which it is to be controlled, one which minimizes heat loss and avoids smoke formation. An advantage is that its optimum setpoint does not need to be adjusted as a function of load. At low loads, the air-fuel ratio must be raised to keep the carbon-monoxide concentration at a constant level, thus raising the oxygen content to essentially the same degree as required by its steam-flow function.

The use of analyzers for optimization in this manner is not uncommon. Yet the control system which operates on that analytical information tends to differ considerably as a function of the economic factors related to the process. In a distillation column, for example, one product usually is controlled at a specified setpoint under all conditions, while the other product composition may be optimized in three possible modes:

1. In the first mode, the energy used to effect the separation is costly relative to the value of product lost by incomplete separation; then, to minimize energy consumption, product compositions are controlled at the maximum impurity levels which specifications will allow. This condition is described by Zone I in Fig. 6-2; its boundary is established by the specifications placed on product compositions.

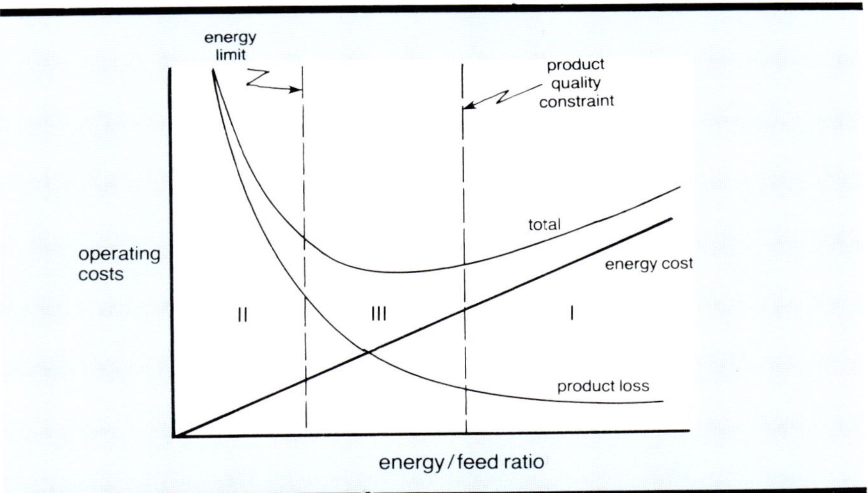

Fig. 6-2. The combination of product loss and energy cost may pass through a minimum.

2. A second operating zone is entered if the cost of the energy is low compared to the value of the specification product (as in the case where the energy is supplied from waste heat). Then the energy input is simply held at a maximum allowable level; product loss is then minimized, although variable with feed rate.

3. A third possibility is presented when product loss and energy cost are comparable. Then there are optimum compositions for one or both products. The optimum compositions are a function of fuel cost, product values, and feed composition, as described in Ref. 1. The optimum setpoints may be calculated

off-line whenever any of these factors changes significantly, and set into the product composition controller.

Loss also may be measured by a flow rate, e.g., as the flow of gas being vented from a vessel under pressure control. Loss of available work can be calculated from the flow and pressure drop across a control valve (2) or inferred from the position of the valve stem. For maximum energy efficiency, i.e., minimum loss in available work, control valves should be driven either fully open or closed (2). Thus valve-position controllers were used in Fig. 4-6 to drive flow-control valves open and in Figs. 2-8 and 4-7 to drive the bypass valves closed. The energy savings thus achieved are not direct but indirect. In Fig. 4-6 less power is used by the compressor because it discharges its flow into less resistance; in Fig. 4-7, lower column pressure increases the relative volatility of the mixture being distilled, thereby requiring less energy to reach the necessary product purities.

A flow measurement can be used in place of valve position when that flow represents a loss. In Fig. 6-3, the flow of steam passing through the reducing valve suffers a loss in available work. The work is recoverable if that steam can be passed through a turbine first. Although it is necessary to control pressure with the reducing valve for quick response to variations in demand, flow controller FC-1 can provide steady-state optimization by decreasing flow to its minimum controllable value. This is accomplished by increasing turbine loading at whatever rate is deemed acceptable. Actually, any process or device that can increase the supply of or reduce the demand for low-pressure steam more profitably than using the reducing valve is a candidate for flow control.

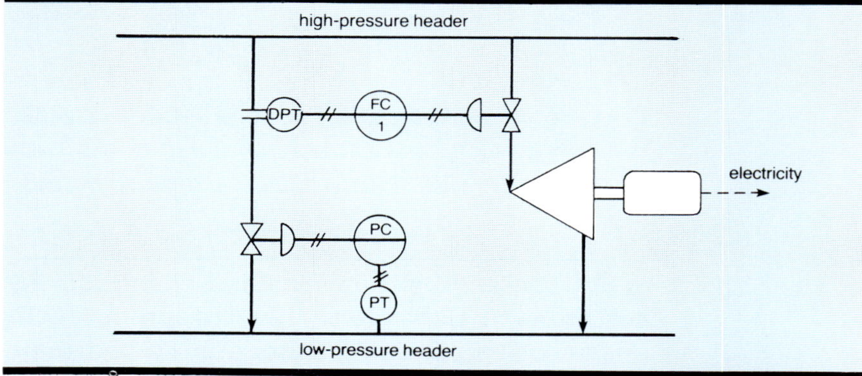

Fig. 6-3. The flow controller gradually adjusts turbine loading to minimize the flow of steam being let down to the low-pressure header.

Note that the systems of Figs. 4-6, 4-7, and 6-3 will perform quite acceptably without the optimizing controller although their efficiency will be less than maximum. The optimizing controller then does not dominate but accommodates, performing its role at a rate selected to minimize disturbance. Optimization also must be confined within acceptable limits so that setpoints will not be driven to unrealistic levels at times when optimization becomes impossible.

6-3. Sequencing by Priority

It is possible to sequence optimizing controllers when there are multiple manipulated variables. For example, when the turbine in Fig. 6-3 reaches full loading, further increases in low-pressure steam demand must be met by flow through the reducing valve. In Fig. 6-4, controller FC-2 begins to decrease steam to the low-pressure feedwater heater when flow through the reducing valve reaches its setpoint, which is higher than that of FC-1.

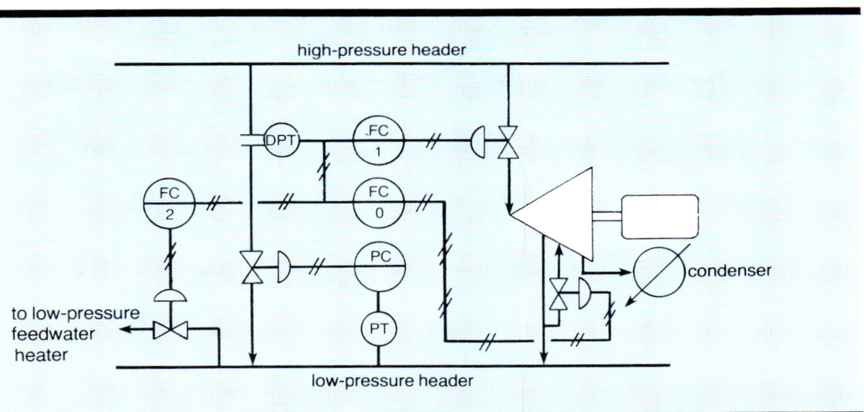

Fig. 6-4. FC-2 closes the feedwater heater on high flow, and FC-0 readmits steam to the condensing section on low flow.

Other controllers may be added in the same manner, with setpoints positioned higher or lower than FC-1, depending on the priority of the manipulated variable. Figure 6-4 also shows FC-0, designated to admit steam to the condensing section of the turbine when demand for low-pressure steam falls below supply.

The difference between the three setpoints need not be great—as little as one percent should be sufficient to keep them from controlling simultaneously. All the optimizing controllers should have only integral action. In this way only one optimizing controller will be active at a time (the others will saturate), and transient disturbances will be countered by the primary controller alone (which is PC in Fig. 6-4).

Figure 6-5 shows two examples of sequencing. Hot and cold water are sequenced to the jacket of a batch reactor, using valve positioners calibrated in split range. The hot-water valve opens above 50% controller output and the cold-water valve opens on a decreasing signal below 50%. A third valve provides chilled water for additional cooling. If the chilled water is added to the jacket as are the hot and cold water, it should be sequenced, as they are, using a calibrated valve positioner. However, in this process chilled water is added through a separate coil having no circulating pump or secondary controller, so that it has no direct effect on jacket-outlet temperature. In this case, the simplest sequencing mechanism is then the valve-position controller acting on the cold-water valve. When that valve approaches full opening (5% signal) the VPC begins to open the chilled-water valve. The resulting additional cooling causes the reactor TC to raise the jacket setpoint enough to keep the cold-water valve in its control range.

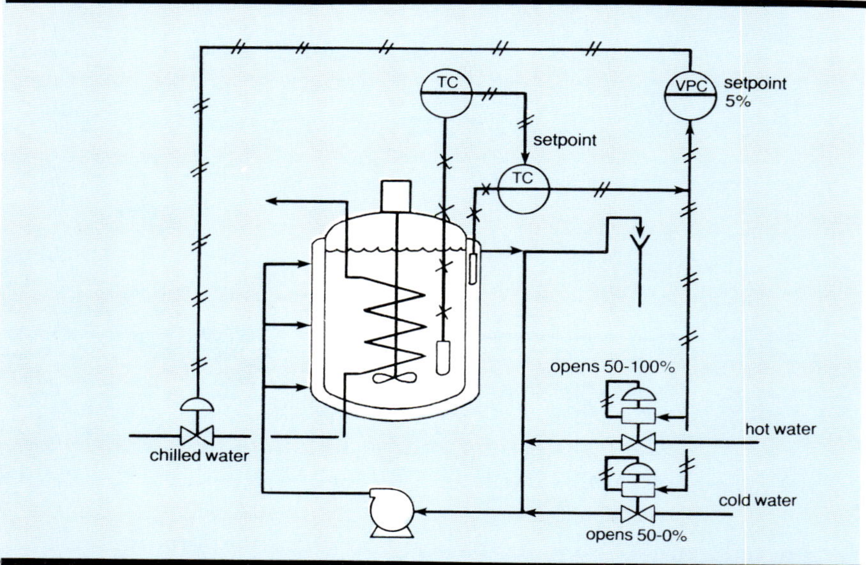

Fig. 6-5. The chilled-water coil is used only when jacket cooling is inadequate.

The VPC is an optimizing controller from the standpoint that the higher-cost chilled water is reserved for use only when the lower-cost cold water is inadequate to meet demand for heat removal.

Because optimizing controllers typically have only integral action, the variables they control, such as flow or valve position, must remain in their manipulable range. In this way the proportional action of the controllers manipulating them will be able to maintain closed-loop stability. Furthermore, sequenced optimizing controllers all must be connected to the same proportional feedback signal, such as the flow measurement in Fig. 6-4. They cannot be cascaded one upon the other or instability will result from accumulating phase lags beyond 180°.

6-4. Hill Climbing

In all of the above examples, the gain of the economic control loop was virtually constant. In other words, the measured loss responded in essentially a linear manner to manipulation. Control was applied near constraints such as 10% and 90% valve positions, at the lowest oxygen level attainable without smoke formation, etc.

There is another class of economic variables which passes through a maximum or minimum as the manipulated variable or variables are adjusted. In fact, it is this variable-gain relationship which directly identifies the optimum condition. The measured variable is in itself an indication of performance, efficiency, or debit, which is to be maximized or minimized as the case may be.

Combustion processes can be optimized in this way: Flame temperature is maximum in a furnace, and knock is minimum in an engine, when fuel and air are blended in proper proportions. Occasionally the products from a chemical reaction will exhibit a minimum value of some physical property, such as density or conductivity, when ingredients are accurately matched.

It's possible to *calculate* certain performance indices which will pass through maxima or minima, such as yield, efficiency, or energy consumption per unit production. While this information may be valuable in guiding operators or informing management about plant conditions, it is rarely amenable to on-line control. Often, the variables which are combined are measured at different times or in different parts of the plant, so that they lack dynamic correlation. It then becomes necessary to dynamically compensate them relative to some reference variable, or, more conveniently, to report values averaged over time. In either case, dynamic response of the control loop is compromised.

The greatest obstacle to controlling at the maximum or minimum, however, is the gain variation itself. The controller must be direct-acting on one side of the maximum (or minimum) and reverse-acting on the other. Furthermore, gain at the optimum is zero, so feedback at that desired operating point is nonexistent. The author describes a slope-seeking controller in Ref. 3, based on dividing the rate of change of the controlled variable by that of the manipulated variable. Again, dynamic compensation is necessary in that the two signals are displaced in time by the response characteristics of the process being controlled.

A severe limitation to this type of controller is that its fundamental control mode is that of integration. It cannot, therefore, be used on processes that lack self-regulation. The 90° phase lags of the controller and process combine to produce undamped oscillations under all conditions of flow and without regard to the time constant of the controller.

It is not surprising, then, that very few hill-climbing controllers have been successfully applied to industrial processes. An alternative usually exists which allows safer and more responsive control action, such as optimizing fuel-air ratio by controlling flue-gas oxygen content instead of flame temperature.

6-5. Programmed Optimization

If we understand thoroughly the process conditions which result in optimum performance, it is possible to coordinate variables in a program which will approach optimality within the accuracy of our information and our representation of the process. This is essentially the procedure followed in setting air flow in direct ratio to fuel flow—combustion is thereby optimized within the limits of flow meter accuracy and variability in fuel quality. We can even include the necessary augmentation of air flow at reduced loads by applying the appropriate characterization either to the setpoint or measurement of the air flow controller. This is commonly done in power plants by modifying the calibration of the square-root extractor applied to the differential-pressure signal from the metering device. This modification is indicated in the air flow control loop of Fig. 6-1. System accuracy is improved by the addition of the oxygen controller correcting the fuel-air ratio by feedback; nonetheless, the bulk of the control effort is provided by the feedforward coordination of fuel and air flows.

This same procedure is followed for most chemical reactors: Ingredients are fed in carefully controlled ratios which represent the conditions necessary for optimum reactor performance. This is not only the case where ratios are set to assure stoichiometry, but also where certain ingredients are supplied in excess to reduce competing side reactions.

Because chemical equilibria and reaction rates are affected by temperature and pressure, ingredient ratios may be changed as a function of these variables in an effort to maintain performance at optimum levels. Or, alternately, temperatures and pressures may be programmed as a function of reactant compositions to achieve the same effect. This technique has been applied to batch reactors, where the composition of the reaction mass changes with time.

For the distillation column described by Zone III of Fig. 6-2, there exists an unconstrained economic optimum. It may be a function of measured product composition, and therefore directly controllable as described earlier. However if it is not, or if the composition measurement is not available, process modeling often can be used to program measurable variables relative to each other in an optimum manner. For the distillation column, an optimum energy-to-feed ratio exists—a function of energy costs, product values, and feed composition. If feed composition is variable but unmeasurable, the average ratio of controlled product flow to feed rate may be used. Ref. 1, pp. 334-339, develops this concept more completely.

6-6. Maximizing Productivity

Production rate is usually set in an arbitrary manner at the feed inlet to the process. If all product-quality, environmental, and inventory control loops are functional, production rate is less than maximum. In fact, maximum production is established at the point where one of these variables can be controlled no longer.

Limits of control are signaled by the approach of a control valve to the end of its travel. For example, when a cooling-water valve reaches full opening, temperature can no longer be controlled and some alternative action must be taken. In Fig. 6-5, a valve-position controller was used to begin the flow of a more costly cooling medium, such as chilled water.

Another approach is to limit production within the capability of the cooling-water system to remove heat. In a continuous reactor, this can be accomplished by using the valve-position controller of Fig. 6-5 to manipulate reactant feed rate. Then feed rate will be maintained as high as the cooling system will allow, regardless of its condition. Fouled heat-transfer surface or a high-temperature water supply automatically will reduce production rate, whereas more favorable conditions will increase it automatically.

There is a fundamental difference in the dynamic response of the valve-position control loop manipulating supplementary cooling and that manipulating reactant feed. A change in heat transfer starts to affect reactor temperature directly, whereas a change in reactant flow must change reactant concentration first—before the rate of reaction and therefore heat evolution can effect temperature. The inclusion of the large secondary lag of reactant concentration within the valve-position loop requires that the controller have proportional and derivative modes as well as the integral mode necessary for valve-position control. Furthermore, the manipulation of feed rate is generally too slow to control reaction temperature without concurrent manipulation of cooling. Therefore, the temperature controller must always operate well within the range of the cooling valve; the 5% setpoint shown in Fig. 6-4 may not be enough margin to guarantee stability.

Production rate in a batch reactor is determined by temperature. To maximize production rate, the valve-position controller acting on the signal to the cooling-water valve could adjust automatically the setpoint for reactor temperature. There is an additional lag associated with the functioning of this valve-position loop as well. When the VPC setpoint is exceeded, the VPC must reduce the reactor setpoint. For temperature to follow the setpoint downward, additional cooling must be provided initially, although a smaller amount will be necessary when the lower setpoint finally is reached. This inverse-response characteristic introduces a phase shift into the loop, equivalent to dead time. Consequently, the valve-position controller requires a setpoint which gives adequate margin for the additional cooling needed to reduce reactor temperature.

Additional examples of maximizing production rate were proposed in Sec. 2-5, Pacing Systems. In that context, production rate was set by a flow controller, with override when the most-open valve in the system reached the setpoint of the VPC (Fig. 2-6). Elimination of the flow controller would force the system to operate at maximum production rate.

In Fig. 2-6, a single VPC served all valves. In a more complex system, where different controlled variables respond differently to production rate, each valve may require its own controller. Consider, for example, the distillation column in Fig. 6-6. At times, the condenser may approach its heat-transfer limit due to weather conditions or excessive boilup. At other times, the steam valve may be driven fully open, due to reboiler fouling or excessive feed rate. In either case, reducing feed rate can alleviate the condition and keep the column operating under both pressure and temperature control. But because the temperature and pressure respond differently to feed rate, the two valve-position controllers may require different mode settings. Proportional as well as integral action may be required here to achieve the necessary speed of response.

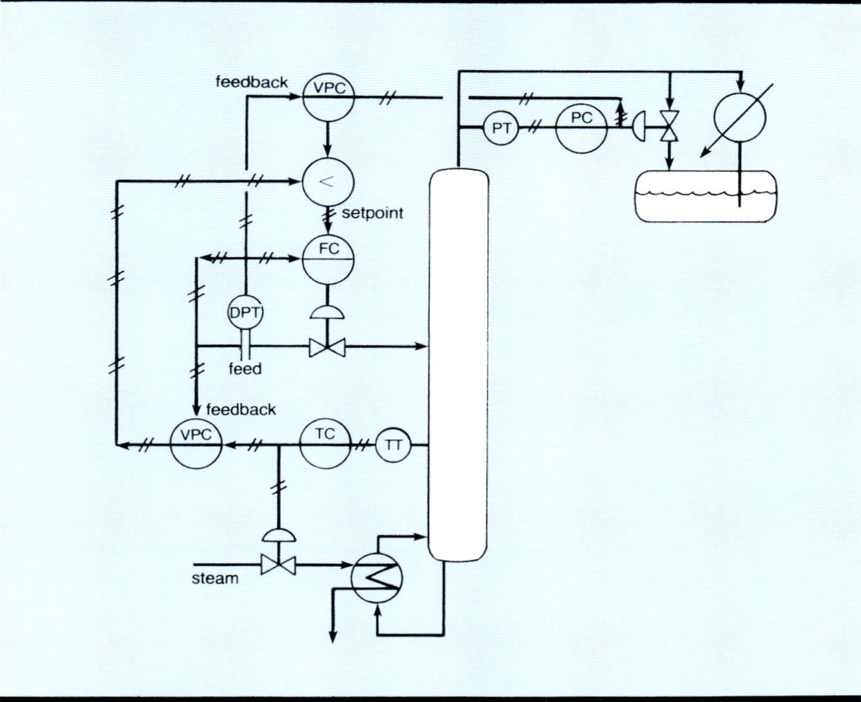

Fig. 6-6. Production rate is limited by the full-load positions of steam and condenser valves.

6-7. Optimizing Multiple Units

When several units are operating in concert, such as boilers producing steam into a common header, their individual contributions to the overall optimum may differ. Then, operating each at its maximum efficiency may not result in maximizing efficiency for the plant as a whole.

Plant optimization is achieved when the total cost of the next increment of product is as low as possible. This condition is coincident with the cost of the next increment of product from all participating units being the same.

Reference 4 describes the optimal allocation of load among four boilers delivering steam to a common header. Each boiler has an efficiency which is a function of its load:

$$E_i = f(L_i) \quad (6-1)$$

The fuel energy required to generate a unit of steam energy is proportional to the efficiency:

$$Q_i = L_i/E_i \quad (6-2)$$

Then the incremental cost in terms of fuel energy to produce the next increment of steam energy is the derivative of Eg. (6−2):

$$\frac{dQ_i}{dL_i} = \frac{1}{E}\left(1 - \frac{L_i}{E_i}\frac{dE_i}{dL_i}\right) \quad (6-3)$$

Incremental cost is seen to be a function not only of efficiency, but also also of its rate of change with load. Therefore, a boiler with low but rising efficiency can have a similar incremental cost to one having a high but falling efficiency. Most boiler-efficiency curves increase with load to a maximum near full load, as shown in Fig. 6-7. As a result, an optimum allocation program starts by giving the first increment of load to the most efficient boiler until its efficiency curve begins to fall. At this point, the next most efficient boiler should be brought on line to share the load with the first. Others are added sequentially on a load-sharing basis.

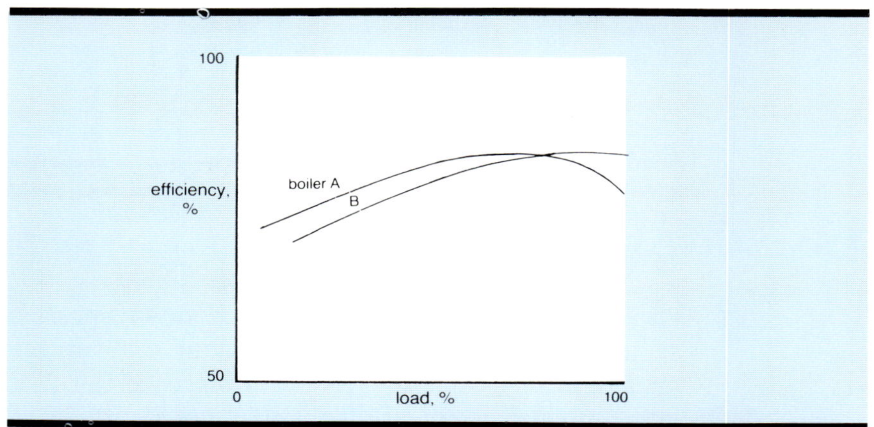

Fig. 6-7. Boiler efficiency typically passes through a maximum near full load.

If the individual efficiency curves are stationary, the corresponding incremental cost curves will be also. Then an optimal load-allocation program can be established and maintained, modified only when one or more boilers is unavailable for service. In actual practice, efficiencies change due to deposition of soot and scale, and variations in heat recovery. Whether the load-allocation program is worth modifying for these contingencies is questionable. Fuel savings resulting from optimal-vs.-equal load allocation are likely to be much less than 1%, so the value of altering the program will be even less.

Savings through optimization at this level are typically meager. A much greater return on less investment usually can be found by optimization at lower levels.

Exercises

6-1. *List some economic variables characteristic of a) chemical reactors, and b) distillation columns.*

6-2. *Which of the above are directly measurable?*

6-3. *Which of them cannot be measured but can be calculated?*

6-4. *Which of them will pass through a maximum or minimum as a result of control action?*

6-5. *What is the most stable method of controlling these variables?*

6-6. *In Fig. 6-7, which boiler has the lower incremental cost at the load where their efficiency curves cross?*

References

[1]Shinskey, F.G. *Distillation Control: for Productivity and Energy Conservation*. New York: McGraw-Hill Book Co., 1977, p. 333.
[2]Shinskey, F.G. *Energy Conservation through Control*. New York: Academic Press, 1978, pp. 26-35.
[3]Shinskey, F.G. *Process-Control Systems*, 2nd Edition. New York: McGraw-Hill Book Co., 1979, p. 162.
[4]Cho, C.H. "Optimum Boiler Load Allocation," *Instrumentation Technology*, V. 25 (October 1978), pp. 55-58.

Unit 7: Control Loop Assignment

UNIT 7

Control Loop Assignment

This unit establishes guidelines for assigning manipulated and controlled variables to form multiple single loops.

Learning Objectives — When you have completed this unit, you should:

A. **Understand the process relationships that cause interactions between variables.**

B. **Be able to estimate relative gains.**

C. **Be able to make single-loop assignments based on relative-gain estimates and dynamic considerations.**

7-1. Open-loop Gains in Interacting Systems

A single control loop is formed by connecting a controlled variable to a manipulated variable through a controller. For that control loop to be both stable and responsive, the control modes of proportional, integral, and derivative must be set in a specific relationship to the steady-state and dynamic characteristics of the process. The product of process and controller gains at the period of oscillation must be less than unity for the loop to be stable, yet, if it is much lower, the loop may not be sufficiently responsive. Therefore, the open-loop gain of the process determines what the controller settings will be.

In a multivariable process, the open-loop gain of a selected controlled variable in response to a given manipulated variable may change when other variables are placed under control. It may have one value when all other loops are open, a different value when one other loop is closed, a third value when another loop is closed, etc.

This concept can be illustrated for a process with two pairs of interacting variables using Fig. 7-1. If loop 2 is opened (m_2 constant) by disabling its controller, the open-loop gain for loop 1 is:

$$\left.\frac{\delta c_1}{\delta m_1}\right|_{m_2} = K_{11}\, g_{11} \tag{7-1}$$

where K represents the steady-state gain and g the dynamic gain vector. Closure of loop 2 forms a parallel path from m_1 to c_1

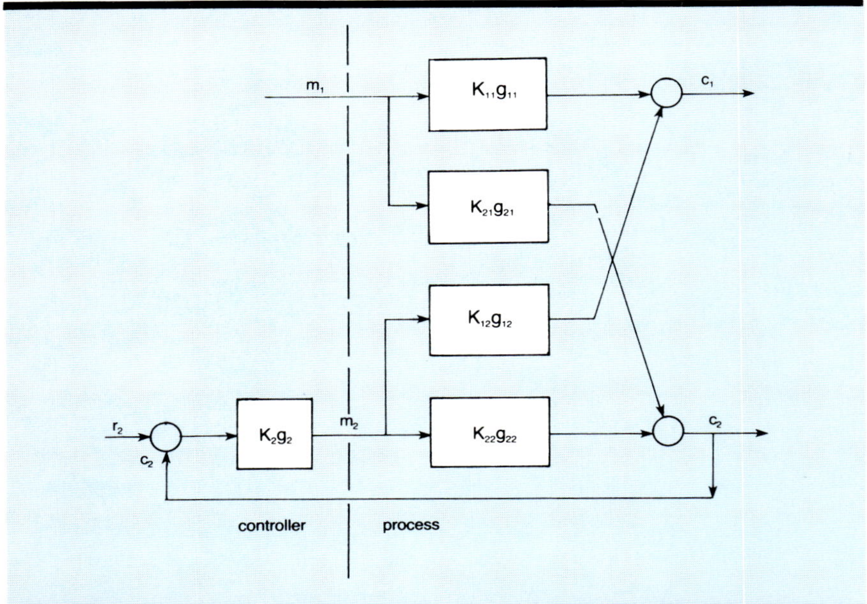

Fig. 7-1. The open-loop gain dc_1/dm_1 changes when loop 2 is closed.

through $K_{21}g_{21}$, loop 2, and $K_{12}g_{12}$. Consider that c_1 is affected by both manipulated variables:

$$c_1 = m_1K_{11}g_{11} + m_2K_{12}g_{12} \tag{7-2}$$

Variable m_2 is driven by its controller

$$m_2 = (r_2 - c_2)K_2g_2 \tag{7-3}$$

which responds to the influence of m_1

$$c_2 = m_2K_{22}g_{22} + m_1K_{21}g_{21} \tag{7-4}$$

Combining the above three equations yields a relationship between c_1 and m_1:

$$c_1 = m_1K_{11}g_{11} + \left(\frac{r_2 - m_1K_{21}g_{21}}{1/K_2g_2 + K_{22}g_{22}}\right)K_{12}g_{12} \tag{7-5}$$

Differentiation of the above expression yields the open-loop gain with controller 2 in automatic:

$$\frac{dc_1}{dm_1} = K_{11}g_{11} - \frac{K_{21}g_{21}K_{12}g_{12}}{1/K_2g_2 + K_{22}g_{22}} \tag{7-6}$$

If either K_{21} or K_{12} approaches zero, the gain of loop 1 is independent of the action of controller 2. The same is true if dynamic gains g_{21} or g_{12} are very low at the period of oscillation of loop 1, which is a function of g_{11}. Loop 1 gain is also a function of the settings of controller 2. As K_2g_2 approaches zero, loop 1 gain will be unaffected by loop 2; at the other extreme, K_2g_2 approaching infinity, Eq. (7-6) approaches:

$$\left.\frac{\delta c_1}{\delta m_1}\right|_{c_2} = K_{11}g_{11} - \frac{K_{21}g_{21}K_{12}g_{12}}{K_{22}g_{22}} \tag{7-7}$$

To illustrate the potential for interaction, consider the case in which all dynamic elements are identical. Then (7-7) reduces to:

$$\left.\frac{\delta c_1}{\delta m_1}\right|_{c_2} = K_{11}g_{11} \quad (1 - K_{21}K_{12}/K_{11}K_{22}) \tag{7-8}$$

This expression isolates the interacting characteristics in a single group. Dividing the open-loop gain for constant m_2 by that obtained for constant c_2 gives the *relative gain* λ_{11}:

$$\lambda_{11} \equiv \frac{(\delta c_1/\delta m_1)m_2}{(\delta c_1/\delta m_1)c_2} = \frac{1}{1 - K_{21}K_{12}/K_{11}K_{22}} \tag{7-9}$$

The response of c_1 to m_1 with c_2 controlled can be shown as a function of the relative gain, by substituting Eq. (7-9) into (7-8):

$$\left.\frac{\delta c_1}{\delta m_1}\right|_{c_2} = \frac{K_{11}g_{11}}{\lambda_{11}} \tag{7-10}$$

If g_{11} can be represented as a first-order lag, the step response of c_1 to m_1 can be illustrated as a function of λ_{11} in Fig. 7-2.

The curves shown are all monotonic. In actual practice K_2g_2 does not approach infinity, so that c_2 is not constant throughout the response of c_1 to m_1. This effect can be qualitatively shown by imposing a delay in the action of controller 2. Then c_1 will respond *initially* to m_1 as if m_2 were constant, but as the *final* state is approached, as if c_2 were constant. This set of curves appears in Fig. 7-3.

For relative gains between zero and unity, loop 2 augments the gain of loop 1, causing it to be more sensitive to control action and slower to respond completely. Relative gains greater than unity preserve dynamic response but reduce the effectiveness of control action by reducing loop gain. Very high relative gains indicate that control will be very ineffective.

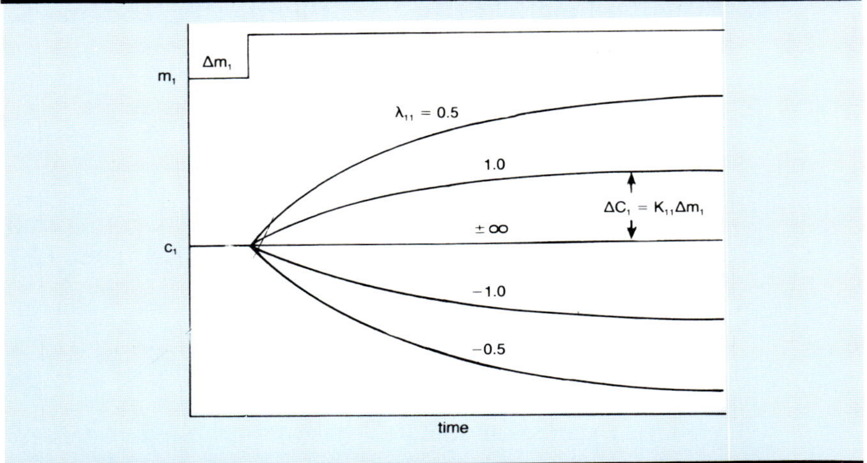

Fig. 7-2. Step response curves for constant c_2 vary in magnitude and direction as a function of relative gain λ_{11}.

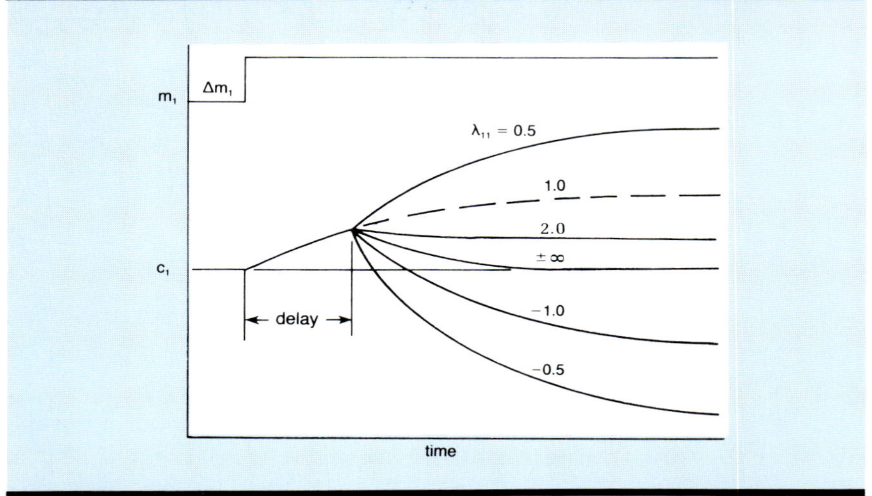

Fig. 7-3. Any delay, which is natural to the response of loop 2, causes a phase reversal to develop when relative gain is negative.

Negative relative gains present a special case. Because loop 1 gain changes sign when controller 2 is transferred between manual and automatic, the action of controller 1 will be incorrect for one of those conditions. If it is set for negative feedback when controller 2 is controlling, it will produce positive feedback when controller

2 is unable to control, either because it is in manual, or because its manipulated variable is constrained. Consequently, loop 1 then is conditionally stable. But there is a dynamic penalty as well—Fig. 7-3 shows that response curves having negative values of relative gain cross through their initial positions on the way to steady state. This is called *inverse response;* the time elapsed before crossover is reached is effectively dead time (Ref. 1). Therefore, even if conditional stability is accepted, dynamic response will be extremely poor.

7-2. Relative-Gain Arrays

To this point, only one relative-gain term has been derived: λ_{11}. In the 2 x 2 system, there will be four, arranged in a matrix:

	m_1	m_2	
c_1	λ_{11}	λ_{12}	1.0
c_2	λ_{21}	λ_{22}	1.0
	1.0	1.0	

(7-11)

The relative gain array has a unique property that all rows and columns add to unity. Then, for the 2 x 2 system, $\lambda_{22} = \lambda_{11}$, and $\lambda_{21} = \lambda_{12} = 1 - \lambda_{11}$. A minimum of $(n-1)^2$ relative gains then need to be independently evaluated to fill an n x n array.

The 2 x 2 array is filled with complementary variables. If $\lambda_{11} = 1.0$, λ_{22} will also, and the other elements will be zero. (This describes the case of incomplete interaction—it will be found when *either* K_{12} *or* K_{21} is zero.) Then m_1 should control c_1, and m_2, c_2. In the opposite situation, given $\lambda_{12} = 1.0$, m_2 should control c_1, and m_1, c_2.

Relative gains in the 0-1 range indicate moderate interaction, with values of 0.5 being the worst. To minimize interaction, variables should be paired whose relative gains are closest to unity. When all four are 0.5, there is no preference, and dynamic properties must be examined to determine the best choice.

When a relative gain exceeds unity, there will appear in the same row and in the same column other relative gains that are negative. Variable pairs having negative gains should not be selected for control for reasons given earlier. The only other choice in a 2 x 2 system (and typically in larger systems as well) is to pair variables having high relative gains. When relative gain exceeds 5, loop interaction becomes severe, and it may not be possible to control both loops at the same time.

7-3. Estimating Relative Gains

Equation (7-9) actually shows two methods of calculating relative gains. The first is the ratio of two gains:

$$\lambda_{11} = \frac{(\delta c_1/\delta m_1)m_2}{(\delta c_1/\delta m_1)c_2} \tag{7-12}$$

and the second requires four:

$$\lambda_{11} = \frac{1}{1 - K_{21}K_{12}/K_{11}K_{22}} \tag{7-13}$$

The first method can be implemented either by testing an existing process or by differentiating a mathematical model of the process.

For the test, m_1 can be stepped with controller 2 in manual, and the steady-state gain $(\Delta c_1/\Delta m_1)m_2$ observed. Then controller 2 is placed in automatic and the step repeated, again observing the steady-state gain, which is $(\Delta c_1/\Delta m_1)c_2$. Their ratio is then calculated.

Differentiating a mathematical model of the process has the advantage of not requiring the test, or even the actual plant, and producing a solution in terms of the variables themselves.

To illustrate, consider the case of mixing two streams in a continuous process to form a third stream of controlled flow and composition. Streams A and B are combined to give a blend of flow F

$$F = A + B \tag{7-14}$$

and a concentration x

$$x = \frac{A}{F} \tag{7-15}$$

The numerator for λ_{FA} is found by differentiating Eq. (7-14) where B is constant:

$$\left.\frac{\delta F}{\delta A}\right|_B = 1 \qquad (7\text{-}16)$$

The denominator is found by differentiating Eq. (7-15) where x is constant:

$$\left.\frac{\delta F}{\delta A}\right|_x = \frac{1}{x} \qquad (7\text{-}17)$$

Then

$$\lambda_{FA} = \frac{1}{1/x} = x \qquad (7\text{-}18)$$

Leaving λ_{FA} in terms of x allows quick estimates of its value under different conditions and also serves as a general solution for all binary blending systems.

In some cases, open-loop gains with other variables under control may not be available. Then, for the 2 x 2 system, Eq. (7-13) can be applied to the four open-loop gains that have been determined with fixed manipulated variables.

The opposite case may also apply, i.e., open-loop gains in terms of the controlled variables alone may be known. These would take the form:

$$m_1 = H_{11}c_1 + H_{12}c_2 \qquad (7\text{-}19)$$

$$m_2 = H_{21}c_1 + H_{22}c_2 \qquad (7\text{-}20)$$

Here coefficient H is the *reciprocal* of the open-loop gain with the other loops under control, i.e.,

$$H_{ij} = \left.\frac{\delta m_i}{\delta c_j}\right|_c \qquad (7\text{-}21)$$

Substitution of Eq. (7-20) into (7-19) and differentiation will produce corresponding values of K; for example,

$$K_{11} = \frac{1}{H_{11} - H_{21}H_{12}/H_{22}} \qquad (7\text{-}22)$$

Then relative gain is the *product* of K_{ij} and H_{ji}:

$$\lambda_{11} = K_{11}H_{11} = \frac{1}{1 - H_{21}H_{12}/H_{11}H_{22}} \tag{7-23}$$

which is the same relationship as given in (7-13).

One further point should be noted in the calculation of relative gains. Because they are ratios of similar open-loop gains estimated under dissimilar conditions, they are not only dimensionless numbers but also are unaffected by dimensions, nonlinearities, and common factors.

7-4. Dynamic Effects

In deriving the relative-gain relationship, certain simplifying assumptions were used. First, controller gain K_2g_2 in Eq. (7-6) was allowed to approach infinity. This represents a limiting case only—the actual values of K_2 and g_2 depend on all the Ks and gs in the process. If controller 2 has integral action, g_2 will approach infinity in the steady state, so that Eq. (7-7) is valid for steady-state conditions. In the short term, however, the actual open-loop gain for loop 1 will vary between K_{11} and K_{11}/λ_{11} (or $1/H_{11}$). An attempt was made to illustrate this property in Fig. 7-3, where the initial step response of $\Delta m_1K_{11}g_{11}$, was followed by the final response of $\Delta m_1K_{11}/\lambda_{11}$ (or $\Delta m_1/H_{11}$).

The second assumption was that all dynamic gain vectors were equal, allowing the elimination of g_{21}, g_{12}, and g_{22} from Eq. (7-7). This is not likely to be true in general, although there will be instances where some degree of cancellation will occur.

Consider, for example, the blending of A and B into a tank where level is controlled instead of flow (Fig. 7-4). Liquid level is a measure of volume V of liquid stored in the vessel, whose rate of change varies with inflow and outflow:

$$\frac{dV}{dt} = A + B - F \tag{7-24}$$

A material balance on component x alone gives

$$\frac{d(Vx)}{dt} = A - Fx \tag{7-25}$$

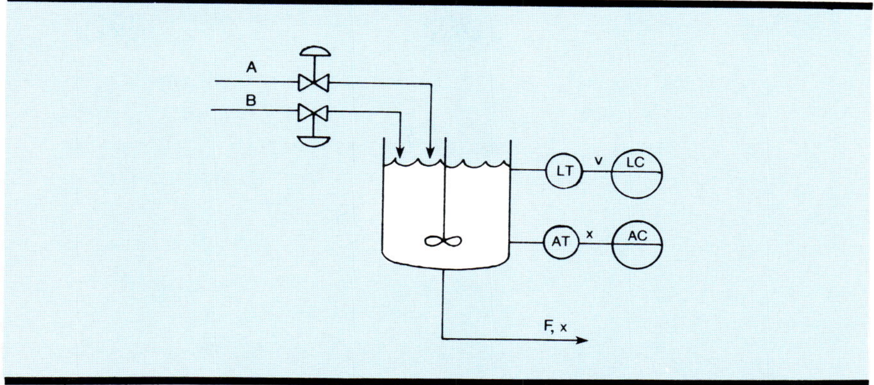

Fig. 7-4. Both liquid level and composition are to be controlled in this vessel.

While neither of these relationships has a steady-state gain $\delta V/\delta A$, their time derivatives do:

$$\left.\frac{\delta\,(dV/dt)}{\delta A}\right|_{B,F} = 1 \qquad (7\text{-}26)$$

$$\left.\frac{\delta\,(dVx/dt)}{\delta A}\right|_{F,x} = 1 \qquad (7\text{-}27)$$

Because x is constant in (7-27), it may be removed from the derivative

$$\left.\frac{\delta\,(dV/dt)}{\delta A}\right|_{F,x} = \frac{1}{x} \qquad (7\text{-}28)$$

Because the differential operator d/dt appears in both numerator and denominator of the relative gain calculation, it has no effect on λ_{VA}:

$$\frac{(\delta V/\delta A)_{B,F}}{(\delta V/\delta A)_{F,x}} = \frac{dx/dt}{d/dt} = x \qquad (7\text{-}29)$$

Where one loop is much faster than the other, interaction is not mutually effective. Consider that the liquid-level loop in Fig. 7-4 will tend to be much faster than the composition loop, whose response is delayed by mixing and sampling times. Then any change made in A (or B) by the slowly acting composition controller can only slightly disturb liquid level, whose faster controller quickly moves B (or A) an equal amount in the opposite direction. However, should a change in outflow F disturb liquid level, its controller in manipulating only B (or A) will cause a profound upset in composition, which will be corrected only gradually by the slower composition controller. Returning to the block diagram of Fig. 7-1, this situation would be described by short time constants for g_{11} and g_{12}, with c_1 being the level measurement, and longer time constants for g_{21} and g_{22}, with c_2 being the composition measurement.

Still another possibility is the existence of short time constants for g_{11} and g_{22}, and longer ones for g_{12} and g_{21}. This arrangement attenuates interaction and may allow loop assignments where steady-state gains are unfavorable, e.g., where $\lambda_{11} < 0.5$.

Occasionally a process will be encountered having a proposed loop with no steady-state gain but considerable dynamic response when other loops are open. However, the closure of other loops may give it a finite steady-state gain. The relative gain for this proposed loop is zero because its numerator is zero. Yet a configuration closing that loop may be effective as long as the other loops are in automatic and retain control. This configuration always would be stable, but only conditionally effective. This application is discussed in more detail in Ref. 2.

7-5. Controller Adjustment

When interaction exists between control loops, each controller moves more than one valve. Equation (7-6) has a second term including all the interacting elements, which augments the open-loop gain of loop 1. If there are an odd number of negative signs among those interacting elements (K_{21}, K_{12}, K_{22}), the loop gain will be increased by interaction, and λ_{11} will fall between zero and unity. Furthermore, the interacting contribution to loop gain will be delayed by the four dynamic elements (g_{21}, g_{12}, g_2, and g_{22}). As a result, controller 1 will need a wider proportional band (lower gain) and longer integral and derivative settings when controller 2 is in automatic than when it is in manual. The same can be said for controller 2, although there may be a difference in degree.

In the system of Fig. 7-4, the liquid-level loop is much faster than the composition loop. When the level controller moves valve B, valve A does not move until composition begins to change, and then comparatively slowly. Therefore, the settings of the level controller are not significantly affected by the interaction. By contrast, the composition controller moves valve A directly, but causes valve B to move at almost the same time through interaction with the fast level controller. Thus, the settings of the composition controller must accommodate the motion of *both* valves. The amount of gain change required for restabilizing the composition loop when the level controller is placed in automatic is equal to the relative gain. If λ_{xA} is 0.6, for example, the composition controller gain must be reduced by 0.6. Because of the great difference in dynamic response of these two loops, neither period of oscillation is likely to change significantly due to interaction. If the loops had similar dynamics, both controllers would require that both proportional band and time constants be doubled. Recognize, however, that one controller could be greatly detuned, allowing the other to retain its original settings. Another possibility involves adjusting one controller for a tight proportional band and slow integral time, with a wider band and faster time for the other, thereby reducing interaction by forcing the periods of the two loops apart.

If the interacting terms of Eq. (7-6) have an even number of negative signs, they will cause a net reduction in the gain of loop 1, although lagged in phase. Relative gain will then exceed unity or be negative. The short-term response of loop 1 is unaffected by this interaction, so its proportional stability is essentially the same whether the other controller is in manual or automatic. However, the loss in gain as steady state is approached requires a much greater control effort than the initial responsiveness will permit, resulting in a general loss of control effectiveness with increasing relative gain. Reference 3 reports the need to double both proportional band and integral time for both temperature controllers on a simulated distillation column where the relative gain of the interacting loops was 5.3.

7-6. Multiple-loop Systems

When more than two loops interact, calculating relative gains becomes more difficult. Matrix methods may have to be used if individual derivatives giving both K_{ij} and H_{ji} are impractical. In effect, a matrix of K gains can be converted into a matrix of H gains by matrix inversion:

$$H = K^{-1} \tag{7-30}$$

Then the matrix Λ of relative gains λ_{ij} is found by an element-by-element multiplication of K_{ij} gains by H_{ji} gains:

$$\lambda_{ij} = K_{ij}H_{ji} \tag{7-31}$$

(Note that the H matrix needs to be transposed because its subscripts are given in a form that is reversed from the K matrix.)

In larger arrays, symmetry disappears and the selection of variable pairs for control-loop assignment becomes more difficult. Consider, for example, the 3 x 3 system given below:

	m_1	m_2	m_3
c_1	0.58	−0.11	0.53
c_2	0	1.0	0
c_3	0.42	0.11	0.47

Making the obvious choices of m_1 to control c_1, and m_2 to control c_2, leaves m_3 to control c_3, although m_3 actually has more effect on c_1. Interaction between the first and third loops in this system will be pronounced.

As more loops make up a system, there will tend to be a greater divergence between those that are fast (flow, level, and pressure) and those that are slow (temperature and composition). Because fast loops are not easily upset by slower loops, it becomes possible to remove them from consideration as being controlled, thereby reducing the number of variables in the relative-gain array.

When fast-responding controlled variables are removed in this way, the manipulated variables arbitrarily assigned to them are removed also. Left in the array are the slow controlled variables and the remaining manipulated variables, constituting a subset of the larger array. The fast controlled variables may later have other manipulated variables arbitrarily assigned to them, in which case the unassigned manipulated variables form a different subset with the unassigned slow controlled variables.

For a process having n manipulated and n controlled variables, there are $(n^2-n)/2$ possible 2 x 2 subsets for two designated (slow) controlled variables. Thus, a 3 x 3 system will have three possible subsets for two designated controlled variables, a 4 x 4 system will have six possible subsets, and a 5 x 5 system will have ten. Not all of the possible subsets will be viable. For example, a distillation column having four manipulated variables will have typically four viable subsets of two composition variables.

Relative gains for the subsets are calculated assuming all of the fast-responding variables are controlled. Thus, equations that are differentiated must not contain any of the arbitrarily assigned manipulated variables because they are not constant.

To illustrate this concept, consider the distillation column of Fig. 7-5. There are four controlled and four manipulated variables shown. The fast controlled variables are pressure p and base level b. Pressure can be controlled by flooding the condenser with reflux R or distillate D, or by throttling steam flow Q. Base level can be controlled by manipulating either bottom-product flow B or steam flow Q. Other combinations are not likely to be effective, due to poor dynamic response.

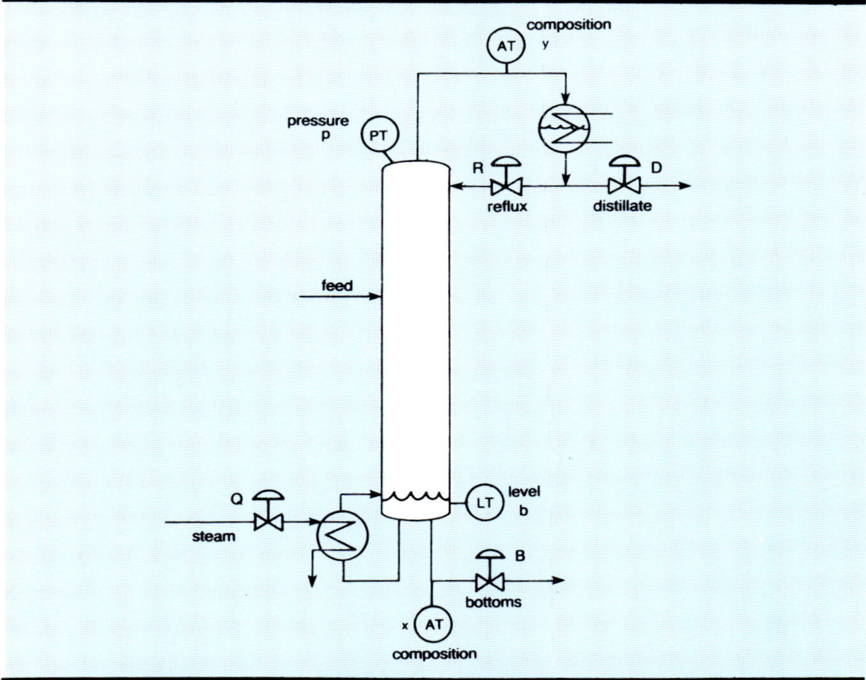

Fig. 7-5. A typical distillation column has four (or more) pairs of variables.

Product compositions x and y may be controlled by any of the four manipulated variables, although B and D may not be manipulated at the same time, because one is dependent on the other and on feed rate.

Considering all of the above limitations, there are only four basic subsets of x and y with different combinations of manipulated variables:

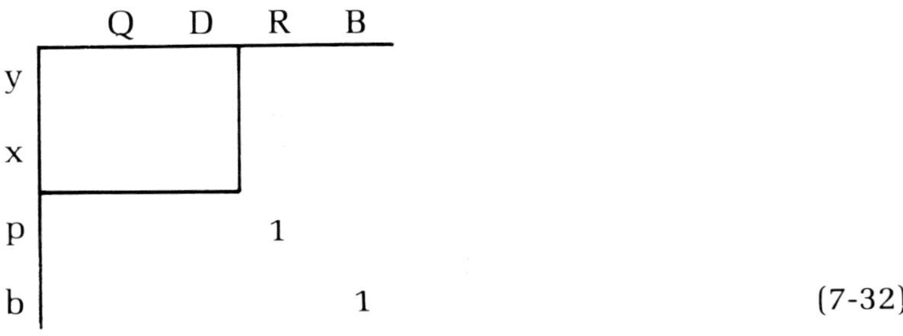

(7-32)

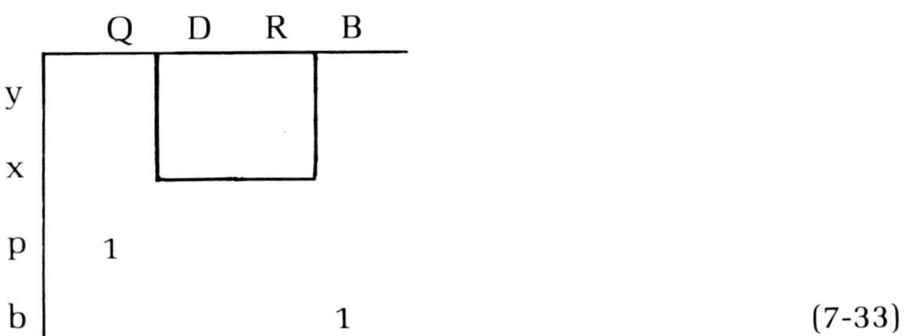

(7-33)

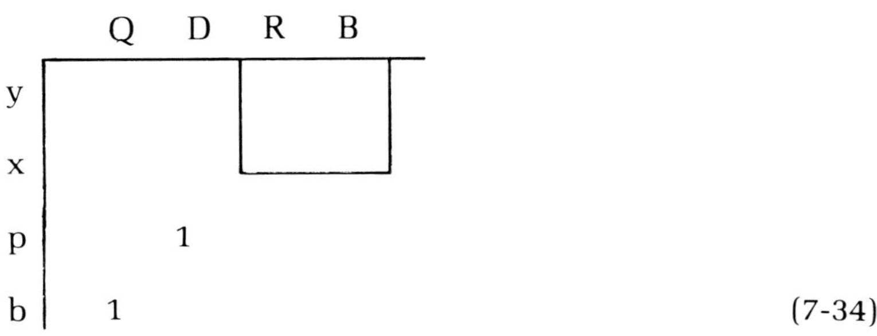

(7-34)

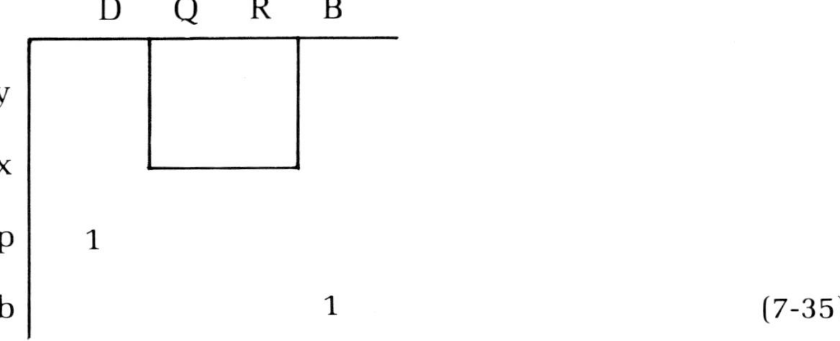

(7-35)

The first three x-y subsets will always contain relative gains in the 0-1 range. Furthermore, λ_{yD} in the D-Q subset will have almost the same value as in the D-R subset, and as λ_{yB} in the B-R subset (4). This relationship facilitates the search for more favorable subsets when the first one examined is unfavorable. For example, if λ_{yD} in the D-Q subset is 0.1, D should not be used to control y. However, D should not be used to control x in the same subset, because of poor dynamic response, even though λ_{xD} is 0.9. Based on the above relationships, however, λ_{xB} in the B-R subset will also be about 0.9, and the dynamic response of this loop should be acceptable.

One further consideration is necessary, having to do with the assignment of the fast loops. Pressure and liquid level are controlled better by large rather than small flows. Pressure is affected by D, R, and Q; but if $D << R$, it will not control pressure as effectively as R or Q. If this is the case in the previous example, then results may be improved by the use of the D-R subset to control x and y, allowing Q and B to control p and b, respectively. The same relationship should be considered in control of base level—if $B << Q$, it may not be effective in that role.

The last of the four subsets is fundamentally different from the others. Because Q drives both x and y in the same direction, and R drives both in the opposite direction, λ_{yR} is always greater than unity. Where it is 3 or less, control may be acceptable, but higher values indicate that another subset should be used.

Exercises:

7-1. *Hot water at 80°C is mixed with cold water at 10°C to produce tempered water at 50°C controlled temperature and flow. Fill out the relative-gain array for this system.*

7-2. *Derive a formula for calculating one relative gain for this system as a function of temperature.*

7-3. *Derive a formula for calculating one relative gain for this system as a function of valve size.*

7-4. *Two streams of the same size are drawing flow from the discharge of a common centrifugal pump; both are under flow control. Estimate a likely range for the relative gain of one flow with respect to its own valve.*

7-5. *Select the most favorable configuration for the process described by the following array of relative gains:*

	m_1	m_2	m_3
c_1	1.6	0.6	−1.2
c_2	0.4	0.2	0.4
c_3	−1.0	0.2	1.8

7-6. *A distillation column of the type shown in Fig. 7-5 has the following conditions:* $D = 0.8$, $B = 0.2$, $R = 4$, $Q = 4.8$; *for the D-Q subset,* $\lambda_{yD} = 0.2$, *and for the Q-R subset,* $\lambda_{yR} = 7.8$. *Choose the best configuration for the four single loops.*

References

[1]Shinskey, F. G. *Process Control Systems*, 2nd Ed. New York: McGraw-Hill Book Co., 1979, pp. 236, 237.
[2]*Ibid.*, pp. 210-212, 220,221.
[3]Toijala, K. and K. Fagervik. "A Digital Simulation Study of Two-Point Feedback Control of Distillation Columns," *Kemian Teollisuus* V. 29 (January 1972).
[4]Shinskey, F. G. *Distillation Control: for Productivity and Energy Conservation*. New York: McGraw-Hill Book Co., 1978, pp. 306-310.

Unit 8:
Coordinating Multiple Loops

UNIT 8

Coordinating Multiple Loops

This unit describes methods used to reduce unfavorable interactions between control loops operating on a multivariable process.

Learning Objectives — When you have completed this unit you should:

A. Know how to coordinate control loops to reduce interactions.

B. Be able to evaluate the effectiveness of a coordinated system.

C. Appreciate the operational features of coordinated systems.

8-1. Using Meaningful Variables

Occasionally the engineer will have a choice about which variables should be controlled to achieve the desired process objective. This opportunity arises particularly when a true measurement of product quality is not available and another must be substituted.

A case in point is the control of environmental conditions in an area where fibrous products are dried or stored. Both temperature and humidity need to be controlled, but many possible measurements are presented. Fig. 8-1 is a psychrometric chart, wherein absolute humidity of air is plotted against dry-bulb temperature. But the chart contains other coordinates as well: Wet-bulb temperature, relative humidity, and dew-point temperature. Controlling any two of these variables will fix the environmental conditions in the space. For example, controlling relative humidity and wet-bulb temperature can accomplish the same purpose as controlling dry-bulb temperature and absolute humidity or dew-point temperature.

Some measurements may be made more easily or more accurately than others, and these factors will affect the choice. For example, dry-bulb temperature is a much simpler and more reliable measurement than wet-bulb temperature; similarly, dew-point

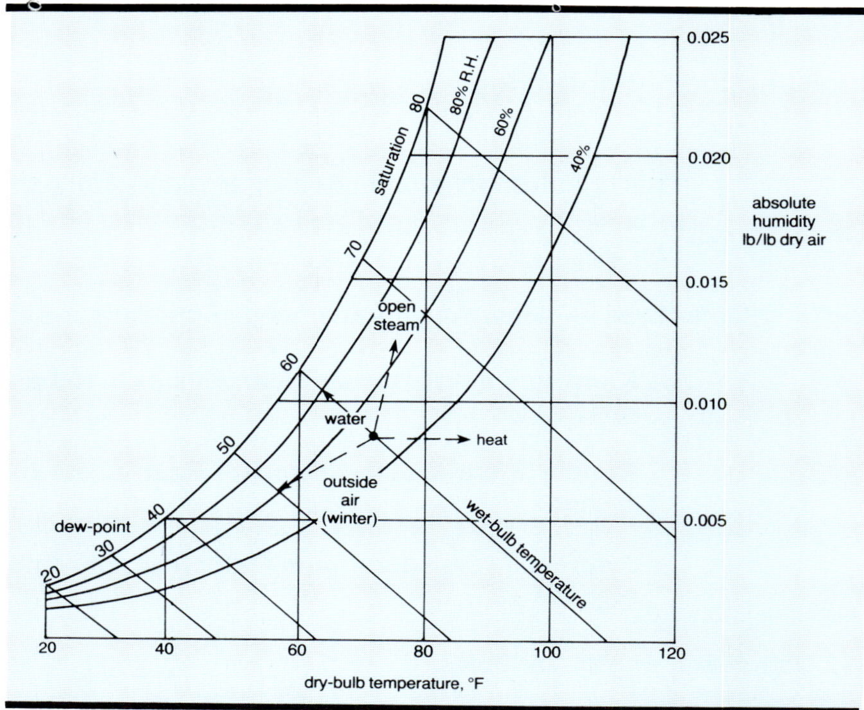

Fig. 8-1. The interaction between temperature and humidity controls is a function of the angle between the manipulated vector and the controlled variable scale.

temperature probably can be measured more accurately than relative humidity and more easily than absolute humidity. But the objective of the process must be the prime consideration. If the environment needs to be controlled to insure the dimensional stability of fine paper stored there, relative humidity may be the most representative index of that property.

The interaction which develops between the control loops is as much a function of the choice of manipulated variables as that of controlled variables. Fig. 8-1 shows vectors representing four possible manipulated variables which might be used to control the environmental conditions. Indirect heating and cooling (without condensation) will not affect absolute humidity or dew-point temperature, thereby producing vectors at right angles to the dry-bulb temperature lines. They therefore may be used to control dry-bulb temperature without upsetting a dew-point or absolute-humidity controller. Water injection has no effect on wet-bulb temperature, but its vector crosses relative-humidity contours at nearly right angles. Open steam may be used to control absolute humidity or dew-point with minimal effect on dry-bulb temperature.

One needs to be careful when using outside air as a manipulated variable. Its vector points to the temperature and humidity of the outside air, which changes seasonally, and to a certain extent, daily. During wintertime it can cool and reduce absolute humidity with little effect on relative humidity. At other times of the year its angle is more variable, making it less reliable for control purposes.

8-2. Combining Variables

There is no need to be limited to single measurable or manipulable variables. If a more meaningful variable happens to be a mathematical combination of two or more measurable or manipulable variables, there is no reason why it cannot be used.

Calculations often are made on inputs to controllers. For example, the tail gas from a Claus sulfur plant contains both hydrogen sulfide and sulfur dioxide. The plant will function most efficiently if these components are maintained in a 2:1 ratio by adjustment of the air-to-feed ratio. Fig. 8-2 shows a divider calculating the H_2S/SO_2 ratio, and a multiplier converting the air/feed ratio into an air flow setpoint.

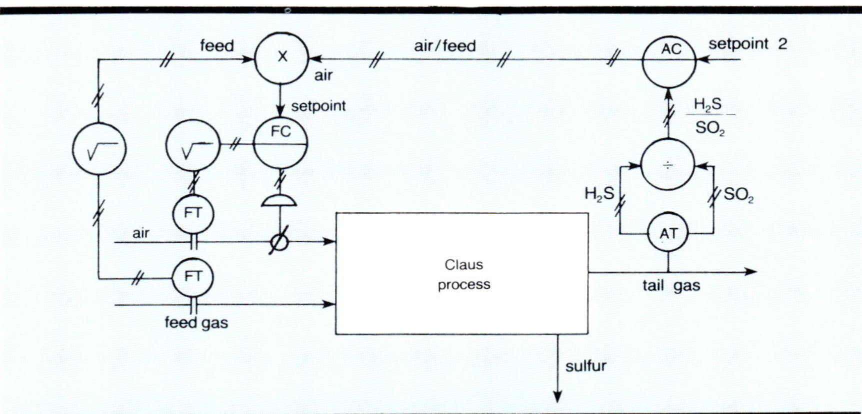

Fig. 8-2. Calculations are made on both input and output of the composition controller.

Mathematically derived *controlled* variables are generated using forward calculations with measurements as inputs. However, mathematically derived *manipulated* variables require backward calculations.

Reference 1 describes a distillation control system wherein the author elects to manipulate the sum of reflux plus distillate for pressure control, and the ratio of distillate to that sum for overhead composition control. His reasoning is that column pressure is affected equally by both streams, so manipulating their sum eliminates a source of interaction. Similarly, manipulating a ratio for composition control moderates the interaction observed when manipulating distillate or reflux flow alone. His control loop arrangement is shown in Fig. 8-3.

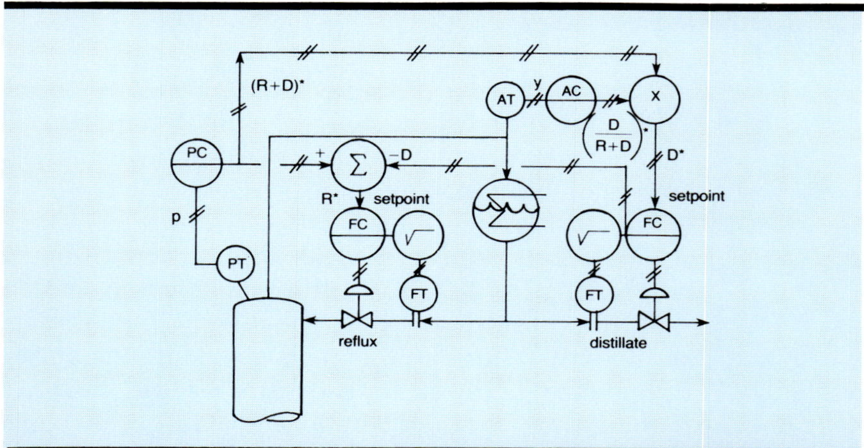

Fig. 8-3. The use of calculating devices allows controllers to manipulate mathematic combinations of variables.

In order for the output of the pressure controller to represent (R + D)*, it must input to a subtractor where D (or R) is subtracted to produce the setpoint R* (or D*). (The asterisk is used here to distinguish setpoints or desired values from measured or actual values.) Similarly, sending (R + D)* to a multiplier generating D* causes the other input to the multiplier to become the ratio D*/(R + D)* by back calculation.

Pressure is controlled by simultaneously manipulating reflux and distillate in a ratio established by the output of the composition controller. Similarly, a change in the output of the composition controller will adjust the ratio without affecting total condensed flow. The composition controller is free to change D*, but any change in D is converted by the subtractor into an equal and opposite change in R* to avoid upsetting column pressure. This action also improves the dynamic response of the composition loop because changes in D* are converted directly into changes in R* without waiting for the pressure controller to respond.

While this system explicitly decouples the composition controller from the pressure loop, the reverse is not necessarily true. Changes in heat input introduced at the reboiler will cause the pressure controller to respond by changing both R and D in a constant ratio. However, maintaining their ratio constant does not guarantee that y will remain constant. The effectiveness of this strategy is evaluated later in the unit.

8-3. Reversing the Process Model

The most direct method for designing a decoupling system is to reverse the model of the process that was used to estimate relative gains. This procedure is well established in the design of feedforward systems (2) which are but a lower-order form of decoupling.

To illustrate the technique, consider a process having three interacting pairs of variables with the following structure:

$$c_1 - c_2 = f(m_2 - m_1) \tag{8-1}$$

$$c_1 + c_2 = f(m_3) \tag{8-2}$$

$$c_3 = f(m_1 + m_2) \tag{8-3}$$

The control system is to develop individual outputs from a plurality of inputs. Consequently, each manipulated variable is to be found as a function of controlled variables:

$$m_2 - m_1 = f^{-1}(c_1 - c_2) \tag{8-4}$$

$$m_1 + m_2 = f^{-1}(c_3) \tag{8-5}$$

$$m_3 = f^{-1}(c_1 + c_2) \tag{8-6}$$

Equation (8-6) is already a solution since a single manipulated variable is expressed as a combination of controlled variables. Then Eqs. (8-4) and (8-5) must be solved simultaneously to develop singular relationships for m_1 and m_2:

$$m_1 = \frac{f^{-1}(c_3) - f^{-1}(c_1 - c_2)}{2} \tag{8-7}$$

$$m_2 = \frac{f^{-1}(c_3) + f^{-1}(c_1 - c_2)}{2} \tag{8-8}$$

The inverse functions represent controller functions. Then m_3 is the output of a controller whose input is $(c_1 + c_2)$; m_2 is the sum of two controller outputs, one having c_3 as an input, and the other having $(c_1 - c_2)$ as an input; m_1 is developed in the same way, except that the controller outputs are subtracted. Figure 8-4 describes the system structure.

Observe that the structure is determined by the process model rather than being arbitrarily selected by the control-system designer. It is therefore more capable of meeting the process needs than an arbitrary structure could be. The controller manipulating m_3 has a calculation at its input, whereas m_1 and m_2 are generated by calculations on controller outputs. The difference controller, operating on $c_1 - c_2$, performs the subtraction at its input using c_2 as a setpoint for c_1. This can be done without adding an explicit subtraction, because all controllers operate on the difference between two variables—only a remote setpoint is required.

The process described above is the bottom of an actual distillation column having twin kettle reboilers (3). Controlled variables c_1 and c_2 are the two reboiler levels, and c_3 is column temperature; m_1 and m_2 represent heat input to each reboiler and m_3 is the bottom-product valve drawing from both reboilers in parallel. It was impossible to close all three loops with single-loop controllers—two loops could be closed satisfactorily, but closing the third caused all three to cycle. Installation of the system shown in Fig. 8-4 completely stabilized the process.

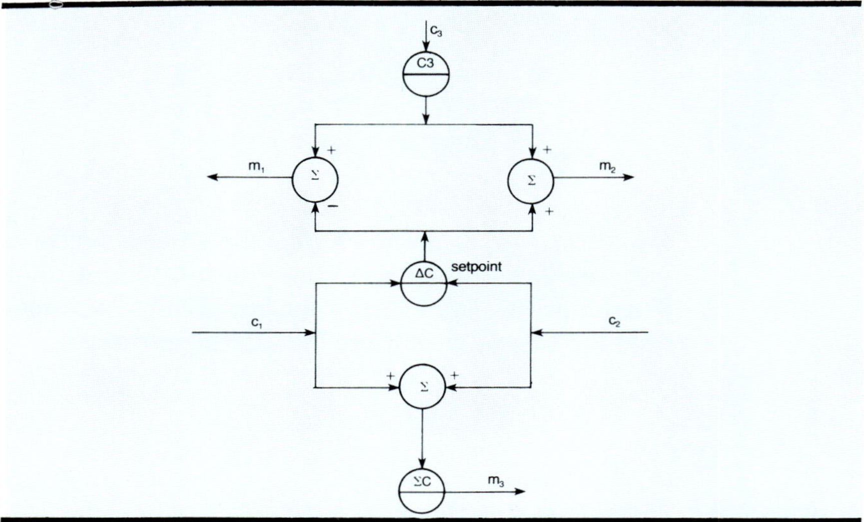

Fig. 8-4. This system can completely decouple the 3×3 process described by Eqs. (8-1) to (8-3).

8-4. Structural Variations

The 2 x 2 interacting process is represented by four blocks, as shown in Fig. 7-1. However, only two blocks are needed for complete decoupling in that only two of the process blocks are responsible for interaction. The obvious and most common structure advocated for decoupling is that of direct combination of controller outputs as shown in Fig. 8-5.

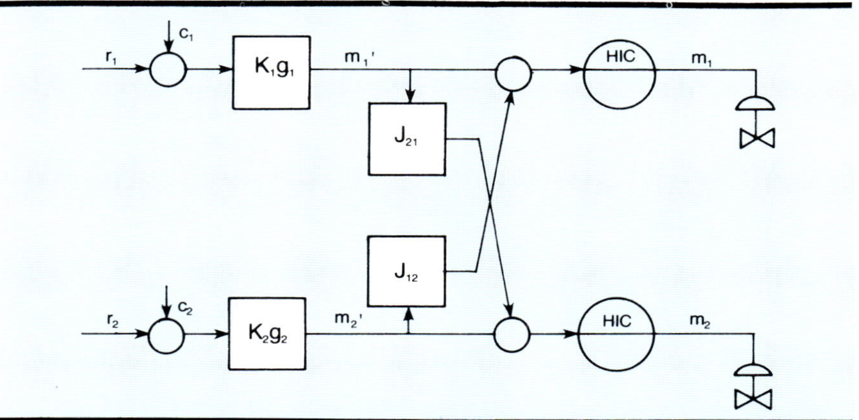

Fig. 8-5. The decoupler which combines controller outputs is difficult to initialize and fails when constrained.

The decoupling functions J_{21} and J_{12} may be derived from the derivatives of the process model:

$$dc_1 = K_{11}g_{11}\delta m_1 + K_{12}g_{12}\delta m_2 \tag{8-9}$$

$$dc_2 = K_{21}g_{21}\delta m_1 + K_{22}g_{22}\delta m_2 \tag{8-10}$$

It is desired to manipulate m_2 as a function of m_1 so that c_2 does not change; then by setting Eq. (8-10) to zero, the gain of the decoupler is found to be:

$$J_{21} = \left.\frac{\delta m_2}{\delta m_1}\right|_{c_2} = -\frac{K_{21}g_{21}}{K_{22}g_{22}} \tag{8-11}$$

The other decoupling function can be found by setting (8-9) to zero:

$$J_{12} = \left.\frac{\delta m_1}{\delta m_2}\right|_{c_1} = -\frac{K_{12}g_{12}}{K_{11}g_{11}} \tag{8-12}$$

With these decouplers in place, the gain of one controlled variable in response to the opposite controller output is zero:

$$\frac{dc_2}{dm_1'} = 0, \frac{dc_1}{dm_2'} = 0 \tag{8-13}$$

However, the open-loop gain of the decoupled loops is not the same as if there were no interaction at all. To determine this relationship consider the half of the system shown in Fig. 8-6. Controlled variable c_1 responds to m_1 directly, and also indirectly, through the decoupler:

$$dc_1 = dm_1' \left(K_{11}g_{11} - \frac{K_{12}g_{12}K_{21}g_{21}}{K_{22}g_{22}}\right) \tag{8-14}$$

or, in terms of dynamic relative gain λ_{11}

$$\frac{dc_1}{dm_1'} = K_{11}g_{11}\left(1 - \frac{K_{12}g_{12}K_{21}g_{21}}{K_{11}g_{11}K_{22}g_{22}}\right) = \frac{K_{11}g_{11}}{\lambda_{11}} \tag{8-15}$$

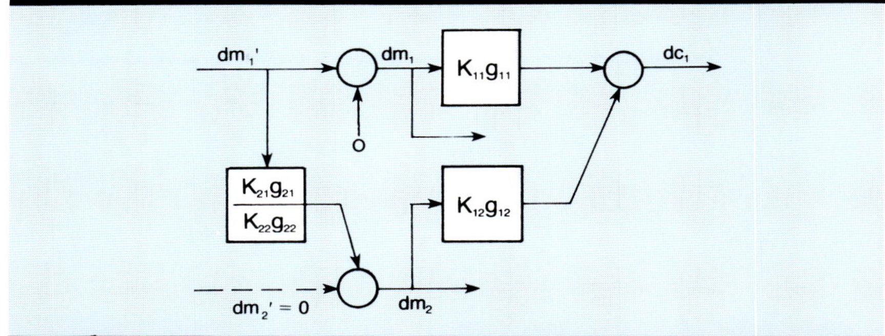

Fig. 8-6. A second, parallel path is formed by each decoupler.

(Note here λ_{11} includes dynamic with steady-state gains.) This is the same open-loop response obtained without decoupling in the case of perfect control of c_2, as described in Eq. (7-10). So while this form of decoupler does eliminate cross-loop interaction, the stability of the decoupled loops is still affected by the interacting elements in the process.

There are two other disadvantages of the structure shown in Fig. 8-5. The controllers cannot be easily initialized, i.e., have their outputs adjusted to initial conditions prior to transfer to automatic. Assume that the control valves themselves are positioned individually at manual (HIC) stations. Having arrived manually at satisfactory valve positions, the operator needs to match controller outputs to those positions to achieve bumpless transfer to automatic. The first controller can be so adjusted, since the input to the decoupler from the other controller is constant. But manually initializing the second controller changes the other manipulated variable through the decoupler, thereby upsetting the loop which is already in automatic.

A second problem appears when a constraint is encountered by one of the manipulated variables. Being no longer free to move, its controlled variable cannot be controlled. However its controller remains in automatic and attempts to control by moving the *other* manipulated variable through the decoupler. When both controllers compete for the single unconstrained manipulated variable both loops will be lost.

Arranging the decouplers in a feedback configuration as in Fig. 8-7 eliminates all three of these problems. Functionally, the decoupler blocks are identical to those shown in Fig. 8-5 because they accomplish the same objective. However they also improve the response of the remaining single loops. To demonstrate, derive the relationship between controller outputs and valve positions in Fig. 8-8.

$$dm_1 = dm_1' - \frac{K_{12}g_{12}}{K_{11}g_{11}} dm_2 \tag{8-16}$$

$$dm_2 = dm_2' - \frac{K_{21}g_{21}}{K_{22}g_{22}} dm_1 \tag{8-17}$$

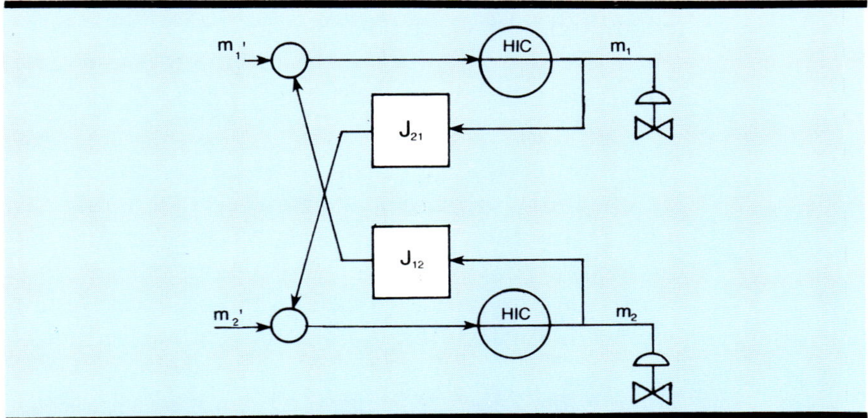

Fig. 8-7. Arranging the decouplers in a feedback configuration eliminates the problems encountered by the structure of Fig. 8-5.

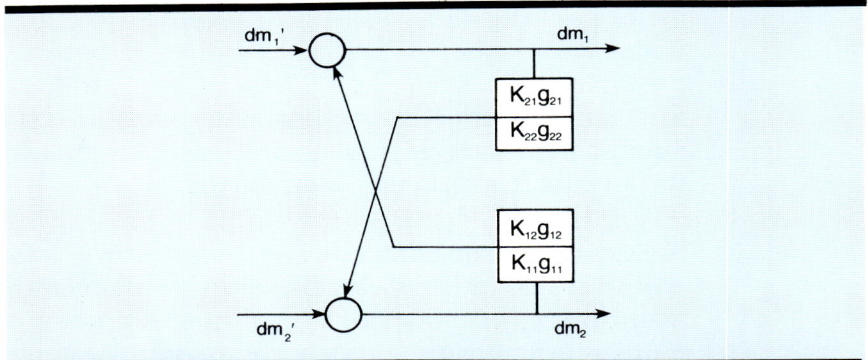

Fig. 8-8. The feedback loop introduces the function λ_{11} between m_1' and m_1.

With dm_2' set to zero, (8-17) may be substituted into (8-16), yielding:

$$\frac{dm_1}{dm_1'} = \frac{1}{1 - \dfrac{K_{12}g_{12}K_{21}g_{21}}{K_{11}g_{11}K_{22}g_{22}}} = \lambda_{11} \tag{8-18}$$

Since $dc_1/dm_1 = K_{11}g_{11}/\lambda_{11}$,

$$\frac{dc_1}{dm_1'} = K_{11}g_{11} \tag{8-19}$$

Initialization of the system in Fig. 8-7 is easily accomplished bumplessly, because controller outputs m_1' and m_2' are back-calculated from valve positions m_1 and m_2 to make each HIC input match its manual output. If in automatic, $m_1 = m_1' + J_{12}m_2$: the back calculation in manual is $m_1' = m_1 - J_{12}m_2$.

Constraints may be imposed either at the input or output of each HIC, as long as the feedback signal is taken downstream so that it reflects the true valve position. In the constrained mode, the feedback signal is fixed, and the unconstrained variable follows it as in a feedforward system. The controller assigned to the constrained variable can then manipulate neither.

The feedback loop through the decouplers must be stable. Its open-loop gain, the product of the two decouplers, must lie between 0 and -1.0 at all periods, for stability:

$$\frac{K_{12}g_{12}K_{21}g_{21}}{K_{11}g_{11}K_{22}g_{22}} = 1 - \frac{1}{\lambda_{11}} \tag{8-20}$$

If the steady-state value of λ_{11}, i.e., $\bar{\lambda}_{11}$, lies between 0.5 and 1.0, the steady-state open-loop gain will fall in the 0 to -1.0 range, so that stability is assured regardless of the dynamic components. If $\bar{\lambda}_{11}$ lies between 0 and 0.5, the steady-state open-loop gain will lie beyond -1.0, so that stability is conditioned on the elements $g_{11}g_{22}$ being faster than $g_{12}g_{21}$. If $\bar{\lambda}_{11}$ exceeds 1.0, the feedback is positive with a gain less than 1.0, and so could be stable given sufficient negative feedback through the control loops—however this is conditioned on the control loops being closed and responsive, which cannot be guaranteed. If $\bar{\lambda}_{11}$ is negative, loop gain is high and positive, and stability is not possible.

8-5. Evaluating Coordinated Systems

Several methods have been proposed to coordinate variables to achieve a degree of control superior to what is possible with uncoordinated single loops. Because the relative-gain function has been useful in evaluating the degree of interaction expected between single loops, it also is applicable in evaluating the interaction that remains after coordination.

For the coordinated system of Fig. 8-3, this amounts to calculating the relative gain of pressure to the sum of reflux and distillate, and that of overhead composition to the ratio D/(R + D). For the pressure loop, results are obvious without calculation—because pressure is affected equally by R and D, manipulating both in parallel eliminates their interaction altogether.

The evaluation of the composition loop requires calculations that are beyond the scope of this text. However, the relative-gain array for the system in Fig. 8-3 is an addition to the four arrays given in Eqs. (7-31) through (7-34):

	$R + D$	Q	$\frac{D}{R + D}$	B
y				
x				
p	1			
b				1

(8-21)

This expands the possibilities in the search for a favorable control-loop structure. By the nature of the coordination imposed, the relative gain $\lambda_{yD/(R+D)}$ will have a value intermediate between λ_{yD} of the D-Q subset and λ_{yR} of the R-Q subset. Since the former is always less than 1.0 and the latter usually much higher, the coordinated system may be more favorable than either.

Those subsets which have Q as one of the manipulated variables for composition control can be compared as to relative gains in the following way:

$$\lambda_{ym} = \lambda_{yR} \frac{dD}{dQ} + \lambda_{yD} \frac{dR}{dQ} \tag{8-22}$$

Since $D + R = Q$, $dR/dQ = 1 - dD/dQ$. When $D/(R + D)$ becomes one manipulated variable, dD/dQ becomes D/Q or $D/(R + D)$:

$$\lambda_{yD/(R+D)} = \lambda_{yR} \left(\frac{D}{R + D} \right) + \lambda_{yD} \left(1 - \frac{D}{R + D} \right) \tag{8-23}$$

Equation (8-23) gives an explicit evaluation of the effectiveness of the coordinated system in terms of the uncoordinated system gains.

For the general case of the feedback decoupler shown in Fig. 8-7,

$$\left. \frac{\delta c_1}{\delta m_1'} \right|_{m_2'} = \frac{K_{11}g_{11} + K_{12}g_{12}J_{21}}{1 - J_{12}J_{21}} \tag{8-24}$$

and

$$\left. \frac{\delta c_1}{\delta m_1'} \right|_{c_2} = K_{11}g_{11} \tag{8-25}$$

The decoupled relative gain is their ratio:

$$\lambda_{11}' = \frac{1 + J_{21}K_{12}g_{12}/K_{11}g_{11}}{1 - J_{12}J_{21}} \tag{8-26}$$

If λ_{11}' is to be 1.0,

$$J_{12} = -K_{12}g_{12}/K_{11}g_{11} \tag{8-27}$$

which was described in Eq. (8-12).

However λ_{11}' can approach infinity if $J_{12}J_{21} = 1.0$.

Therefore, performance is conditioned upon how closely each decoupler represents the true gain ratio of the process elements.

The approach to an infinite value of decoupled relative gain is determined by:

$$1 - J_{12}J_{21} = 1/\lambda_{11} \tag{8-28}$$

As λ_{11} increases in magnitude, $J_{12}J_{21}$ approaches unity and the stability margin of the decoupled system diminishes. Where λ_{11} lies between 0 and 1, the decoupled system always will be stable. When there is a choice of 2 x 2 subsets of a larger array then, as discussed at the close of Unit 7, those subsets having relative gains between 0 and 1 are preferable to those having relative gains exceeding 1, from the standpoint of decoupled stability.

8-6. Partial Decoupling

Using the feedback decoupler of Fig. 8-7:

$$c_1 = \frac{m_1'(K_{11}g_{11} + K_{12}g_{12}J_{21}) + m_2'(K_{12}g_{12} + K_{11}g_{11}J_{12})}{1 - J_{12}J_{21}} \tag{8-29}$$

If J_{21} is zero, and $J_{12} = -K_{12}g_{12}/K_{11}g_{11}$, Eq. (8-29) reduces to $m_1'K_{11}g_{11}$, giving a value of 1.0 for λ_{11}'. Thus relative gain can be corrected using a single decoupler, without the stability risk associated with complete decoupling. Furthermore, complete decoupling is often unnecessary, because one of the controlled variables may be less important than the other or sufficiently rapid in response that it is not upset by the other. Decisions regarding structure are also thereby simplified, in that the systems of Fig. 8-5 and 8-7 reduce to the same structure when one decoupler is eliminated.

An example of a commonly used partial decoupler is the combustion-air flow and furnace-pressure system shown in Fig. 8-9. If the loops are operated independently, each change in air flow will affect furnace pressure. However, the use of the single decoupler, embodied in the dynamic compensator f(t) and the multiplier, eliminates virtually all interaction.

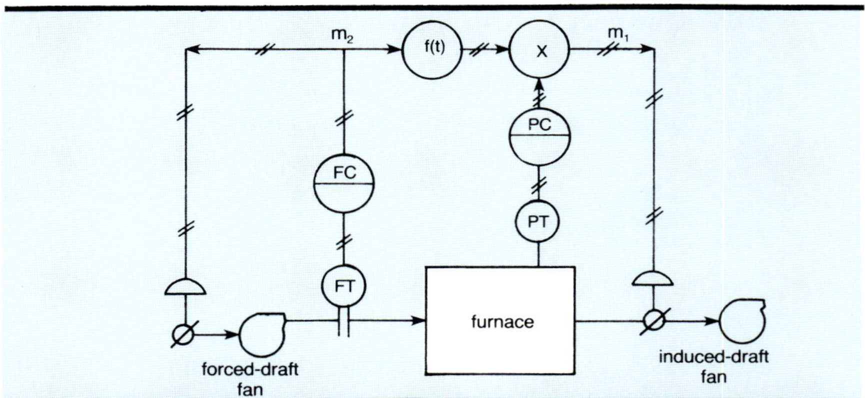

Fig. 8-9. The air flow controller moves both dampers to avoid upsetting furnace pressure when changing flow.

If the multiplier gain is correct, the air flow controller will move both dampers in such a way that flow will change without affecting pressure. If a transient upset appears in pressure, where recovery does not require action by the pressure controller, dynamic compensation is needed in the form of lag or lead-lag characterization.

With accurate decoupling, the pressure measurement will not change, and the pressure controller need not adjust its output (which would upset air flow). Hence, decoupling of flow from the pressure controller is unnecessary.

The application of a single decoupler is exactly the same as implementing feedforward control. For the linear system, the decoupler steady-state gain is a constant (the ratio of two steady-state gains), and its dynamic gain is the ratio of two process dynamic gains:

$$J_{12} = -\left(\frac{K_{12}}{K_{11}}\right)\left(\frac{g_{12}}{g_{11}}\right) \tag{8-30}$$

The dominant dynamic elements in the process are usually first-order lags, in which case their ratio is a lead-lag function. In the case of J_{12}, g_{12} would be the lag and $1/g_{11}$ the lead. Where there is dead time in the process, the dynamic compensator should contain the dead time difference between g_{12} and g_{11}.

Most processes are nonlinear in some respect, and their decouplers should also be nonlinear, representing the process characteristics as closely as practicable. Certain controlled variables are functions of the ratio of two manipulated variables:

$$c_1 = f(m_1/m_2) \tag{8-31}$$

The appropriate decoupling function is found by solving the process equation for the selected manipulated variable:

$$m_1 = m_2 f^{-1}(c_1) \tag{8-32}$$

This requires *multiplying* controller output $f^{-1}(c_1)$ by m_2 rather than adding or subtracting them. The point is that the ratio m_1/m_2 in Eq. (8-31) must be varied to effect a change in c_1. In Eq. (8-32), the ratio is the controller output $f^{-1}(c_1)$. A linear decoupler with a constant gain between m_2 and m_1 will be in error for all conditions except that for which it was adjusted, resulting in ineffective decoupling. In the system described by Eq. (8-32) illustrated in Fig. 8-9, the feedback controller adjusts the multiplier gain, i.e., the ratio m_1/m_2, automatically as process conditions change, thereby keeping the decoupler properly characterized.

Exercises

8-1. *Environmental controls are needed for a room where charts are printed. The printing machines produce more heat than is lost to the surroundings, and the paper absorbs moisture from the air. Select the best pair of controlled and manipulated variables for winter operation.*

8-2. *Given the following process model, develop a control system that provides complete decoupling:*

$$c_1 = K_1 m_1 + K_2 m_2$$

$$c_2 = K_3 m_2/m_1$$

8-3. *Variable c_1 above is flow and c_2 composition of a blend. Due to their pronounced difference in response time, only one variable needs decoupling. Reduce the above design to a partial decoupler.*

8-4. *Using Figs. 8-8 and 8-6, derive Eq. (8-29).*

8-5. *Two of the loops in Fig. 8-4 are structured similarly to those in Fig. 8-5; rearrange the decouplers into a feedback configuration similar to that of Fig. 8-7. Compare your results with Fig. 8-4 from an operational standpoint.*

8-6. *There are a number of partial decouplers (feedforward systems) illustrated in figures in earlier units. Identify them and comment on their linearity and dynamic compensation.*

References

[1]Ryskamp, C. J. "New Strategy Improves Dual Composition Column Control," *Hydrocarbon Processing* V. 60 (June 1980).
[2]Shinskey, F. G. *Process Control Systems*, 2nd Ed. New York: McGraw-Hill Book Co., 1979, pp. 167-171.
[3]Shinskey, F. G. *Distillation Control: for Productivity and Energy Conservation*. New York: McGraw-Hill Book Co., 1977, pp. 303-305.

Unit 9:
Constraint Controls

UNIT 9

Constraint Controls

This unit describes how to place operating constraints on plant equipment, and thereby protect it from entering unsafe or uneconomic operating regions.

Learning Objectives — When you have completed this unit, you should:

A. Know how to bring controllers to and from constrained modes smoothly.

B. Be able to apply multiple constraints to typical control loops.

C. Be able to add safety features that will prevent hazardous conditions from developing even with multiple failures.

9-1. Valve Limits

Loss of control is experienced when a manipulated variable reaches either limit of travel. In most control loops, the limits are the ends of the stroke of a control valve, representing the physical limits of its capacity. However, in certain cases, hard limits must be set short of full stroke to favor equipment constraints downstream of the valve. A valve supplying fuel to a burner, for example, cannot be shut off completely without extinguishing its flame. Therefore fuel valves typically have a low limit which cannot be passed by the controller; shutdown is accomplished with a stop-valve or other on-off mechanism.

When a valve reaches its limit of travel, no matter how that limit is imposed, control is no longer possible. At that point, a deviation may develop at the controller input, with the controller powerless to correct it. If the deviation is sustained, the output of the controller may go well beyond the point of the limit, reaching saturation or "windup." Should the limit then be removed, or the load on the process changed to bring it within the limits of the valve, the controlled variable will recover, typically overshooting the setpoint and requiring time for restabilization. Most controllers have standard or optional "antiwindup" circuits that can be adjusted to inhibit the integral control mode at the point where the limit is reached. Others have internal limits which may be set to stop the valve and also prevent windup.

In a cascade system, the secondary controller can be protected against windup by one of the above methods, but the primary controller cannot, because its output is not the position of the limited valve. Protection requires information on the loss of control by the secondary controller. This can be provided by the system shown in Fig. 9-1.

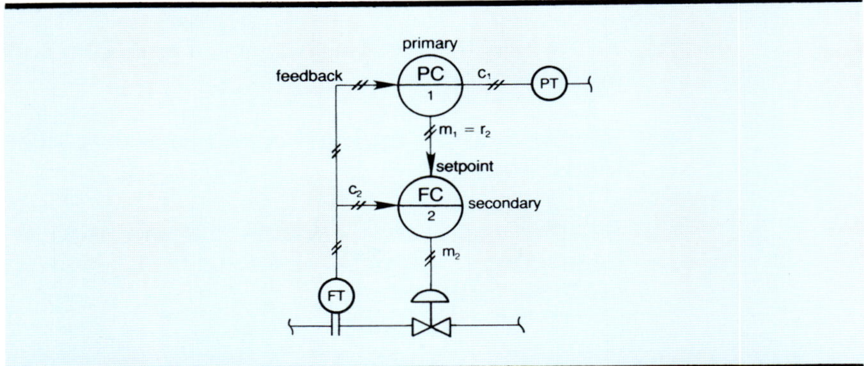

Fig. 9-1. Feedback of the secondary measurement can prevent windup of the primary controller.

In many controllers, integral action is achieved by positive feedback of the controller output through the integral time constant. In Fig. 9-1, this feedback connection is made external and originates from the secondary measurement instead of its setpoint (which is the primary output). If the secondary controller is controlling, these two signals have the same value, and positive feedback will take place, achieving integration in the primary controller. However, failure of the secondary to control will cause a deviation to exist between the primary output and its feedback signal, thereby opening the feedback loop. In this mode, the primary controller loses its integral action; it will be smoothly restored as soon as control of the secondary variable is regained.

For this system to function correctly in the normal operating mode, the secondary controller must have integral action. Note also that the integral time of the primary controller is augmented by the response of the secondary loop. This is, however, a stabilizing feature, particularly if the response of the secondary loop is variable.

In cases where a feedforward calculation is interposed between primary and secondary controllers, as in Fig. 9-2, the system becomes more complex. The output of the primary controller is no longer the secondary setpoint, but is the set ratio of the secondary variable to the feedforward variable. In this case, the true ratio must be calculated from the measurements, for feedback to the primary controller, as shown. The operation and adjustment of this system is identical to the simpler one of Fig. 9-1. In essence, the feedforward calculation must be duplicated (in reverse) in the feedback loop.

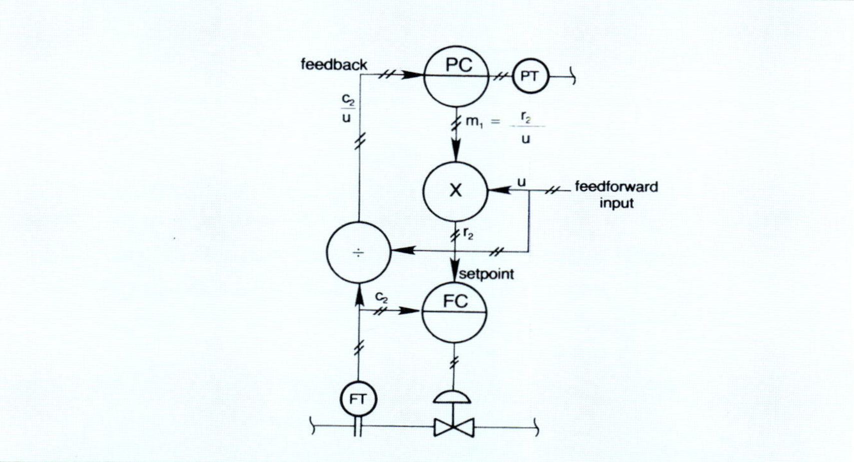

Fig. 9-2. Back calculation of the feedforward function in the primary feedback circuit provides protection from windup.

Another possibility is the formulation of an extremely complex calculation between the output of the primary controller and the setpoint of the secondary. It is not necessary to perform the exact back calculation in the feedback path—a simple summation of secondary deviation with primary output will suffice. Figure 9-3 shows how it is accomplished. As long as secondary deviation is zero, primary feedback will take place, but a non-zero deviation will arrest it.

9-2. Equipment Limits

Most pieces of plant equipment have less operating range than a control valve. It is possible to exceed safe or efficient operating conditions well within the limits of travel of the valve. A certain amount of protection can be provided by limiting the output of the controller to prevent exceeding the acceptable operating range of the equipment, but with some degree of uncertainty. Other factors,

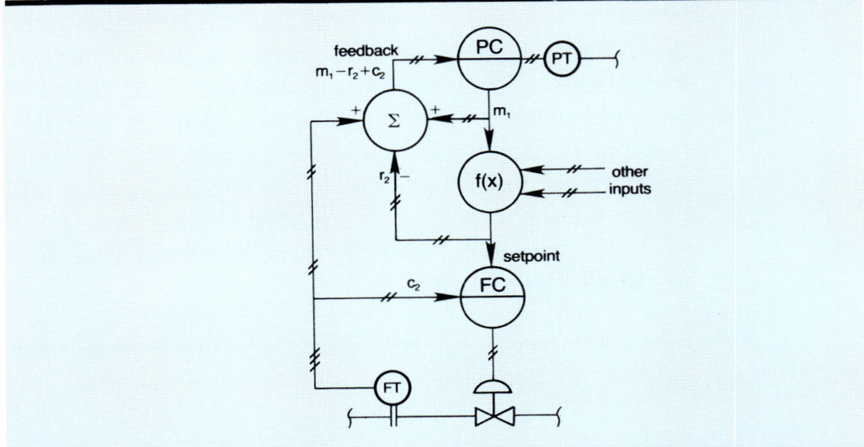

Fig. 9-3. This system applies conditioned feedback to the primary controller to prevent windup for any secondary setpoint calculation.

such as production rate and environmental conditions, typically influence the relationship between valve position and equipment constraints. Trying to protect against all possible unsafe conditions by using valve limits alone may unnecessarily constrain the process.

More accurate protection can be provided by measuring the offending variables or some properties related to them. A controller applied to each measurement can then override the valve should that variable reach its setpoint. This technique is applied to a distillation column in Fig. 9-4 to protect against both excessive and insufficient boilup, as sensed by differential pressure across the column.

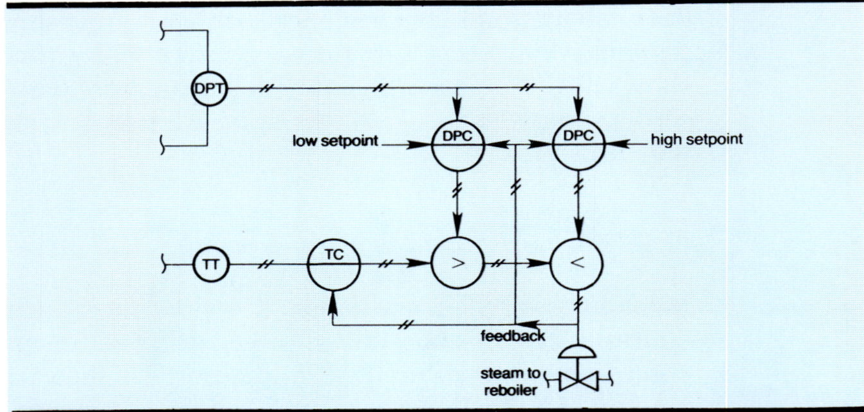

Fig. 9-4. The temperature controller is free to manipulate the steam valve only while the column differential pressure remains between high and low setpoints.

In the normal operating mode, column temperature is controlled by manipulating the steam valve. Upon loss of feed, however, the temperature will tend to rise, causing the controller to close the steam valve. But insufficient vapor velocity through the column will allow liquid to seep through the trays, as indicated by a fall in differential pressure. The DPC with the lower setpoint will respond by raising its output above that of the temperature controller, thereby taking control of the steam valve through the action of the high selector (>).

Similarly, excessive boilup may cause the differential pressure to rise to the point where liquid no longer can flow through the tray downcomers. At the high setpoint, the other DPC reduces steam flow through the low selector before flooding can develop.

It is imperative that all controllers react promptly to an approach to zero deviation—otherwise column performance may deteriorate drastically and affect other operating units. Therefore, windup must be prevented in the controllers which are not controlling. This is accomplished by feedback of the valve signal to all controllers. If there is a secondary flow-control loop as in Fig. 9-1, the flow measurement may be fed back to all primary controllers.

There is no limit to the number of constraint controllers that can be added using selectors as done in Fig. 9-4. Column pressure and base level can be kept in bounds by connecting their controller outputs to the low selector. Recognize, however, that the application of constraint controls in this manner does not solve a problem that originates elsewhere—it only places bounds on it. Any override will cause loss of the normally controlled variable, which is necessarily costly. If its control is to be restored, the cause of the upset must be removed, or production curtailed.

Locating the cause of the override is often difficult, and it may be beyond correction. Frequently it is due to ambient conditions, such as insufficient cooling during hot weather. In cases like this, the only recourse may be to reduce production until the condition abates. Usually this action is taken manually, because a change in production rate may affect several process units, and the dynamic response of the overridden controlled variable to production rate may be quite slow. But in situations where neither of the above objections is significant, it is possible to transfer the overridden controller to a manipulation of production rate. The system capable of changing its structure in this way is covered in more detail in Unit 10, Variable System Structures.

The override controls which use selectors, such as those described in Fig. 9-4, enforce immediate constraints over the process. The differential-pressure controllers on the distillation column prevent that variable from going beyond the limits set by acting on the most responsive manipulated variable—the steam valve. The cost of enforcing these immediate constraints is the loss of temperature control. Another method of achieving differential pressure control, without sacrificing temperature control, would have the DPC manipulating or overriding feed rate. The constraint thus achieved would be delayed in that differential pressure does not respond directly to feed rate, but only indirectly, through the action of the temperature controller in response to feed rate. In this case, differential pressure could exceed setpoint for a considerable time, while the reaction of the temperature controller to the feed rate change was awaited. Eventually the setpoint would be reached, but such response would not be satisfactory for this type of process.

9-3. Multiple Quality Specifications

Some final products must meet or exceed several specifications at once. This is particularly true of products that are distilled: A minimum purity is required, along with maximum concentrations of lower-boiling and higher-boiling impurities. There may be but a single manipulated variable assigned to the product quality controller. Then provision must be made to control at one of the specifications, while the product is better than required on the other two counts.

For example, the distillate product may have to meet or exceed a certain minimum purity specification y_l^m:

$$y_l \geq y_l^m \tag{9-1}$$

Where y represents concentration and subscript l identifies the "light key," which in this case is the dominant component in the distillate product; superscript m identifies the term as being a minimum specification. In practice, dominant components cannot be measured as accurately as those with lesser concentrations. Hence the principal impurities y_{ll} and y_h, representing lower- and higher-boiling components, are measured. The lower-boiling component in the distillate cannot usually be controlled directly;

but is a function of upstream processing; the higher-boiling component or "heavy key" y_h is the controlled variable. Then the setpoint for y_h is generated as:

$$y_h^* \leq 1 - y_{ll} - y_l^m \tag{9-2}$$

There also may be a maximum allowable value of y_h, identified by superscript M:

$$y_h^* \leq (1-y_{ll}-y_l^m),\ y_h^M \tag{9-3}$$

Furthermore, Unit 6 described the possibility of an optimum value of a key impurity in situations where product losses and utility costs were similar in magnitude. The optimum value y_h^o may be less than the other limits. Then y_h^* is selected as the minimum of the three values:

$$y_h^* \leq (1-y_{ll}-y_l^m),\ y_h^M,\ y_h^o \tag{9-3}$$

Figure 9-5 shows how a low selector is used to generate the setpoint.

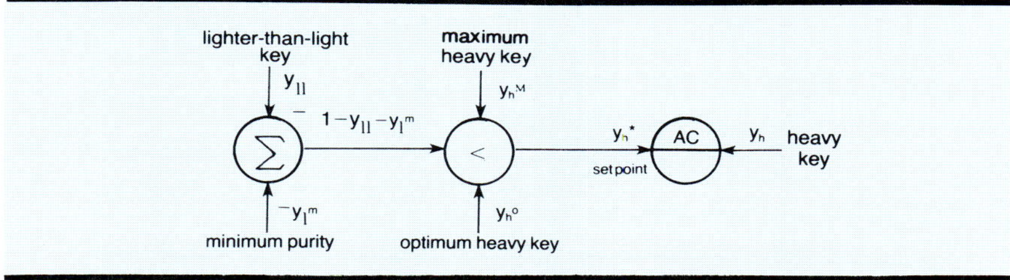

Fig. 9-5. This system uses a signal selector to find the setpoint that will satisfy all requirements.

9-4. Redundant Controls

Certain critical processes need to be protected against instrument as well as equipment failures. This can be accomplished by fabricating redundant instrument channels with selection of the variables closest to safe limits.

An example of this application is the control of furnace pressure for a large, balanced-draft boiler. If the furnace pressure should fall too low, the walls of the furnace could implode, causing extensive damage. If a single pressure transmitter were used, failure of its output in a high direction could cause the pressure controller to react by reducing the pressure to a dangerous level. However, by adding a second pressure transmitter and a low selector, the lower of two pressures would be controlled. Then a single high failure would have no effect on operation, since that signal would be rejected by the selector.

The possibility remains, however, of a single low failure: The failed signal would be selected for control, causing the pressure to be raised to a possibly dangerous level. If this condition is also unacceptable, three transmitters must be used whose outputs are compared in a median selector, as shown in Fig. 9-6. The median selector rejects both the high and low signals, transmitting the remaining one. In the event of a single failure, it selects the remaining two. It also is very effective at filtering noise, rejecting transients in either direction that are not common to at least two of the signals.

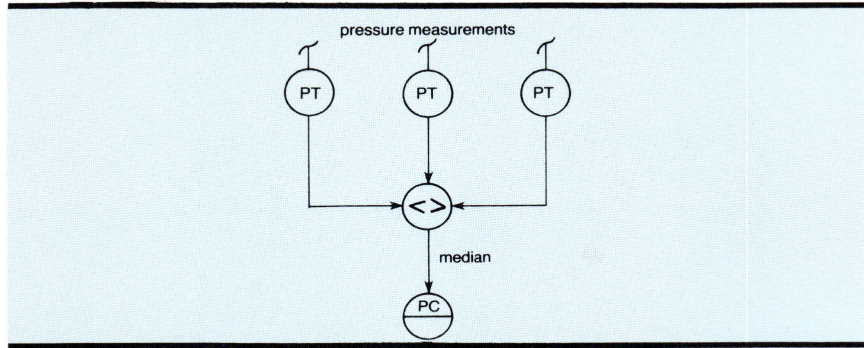

Fig. 9-6. The median selector protects against a single failure in either direction.

Redundancy can be extended to the control function by using a controller for each signal and selecting at their outputs. It even can be extended to the valve by using a valve for each controller and each measurement. If the valves are connected in parallel, the highest flow will be obtained; if they are installed in series, the lowest will be selected. When working with extremely dangerous materials, complete redundancy may be necessary, particularly where valves have a significant failure rate. The backup channel may not need a complete control capability, but only a shutdown capability.

Redundant instruments also can be applied to control a process that is continually changing. A case in point is the fixed-bed catalytic reactor. Due to variations in catalyst activity and production rate, the temperature profile may shift with time. Yet it is necessary to control the hottest temperature to avoid runaway and consequent damage. In this case, multiple temperature sensors may be installed, with the highest selected for control.

9-5. Emergency Controls

This category includes controls that are not used in the normal functioning of the plant, but are activated only in event of an emergency or other abnormal condition, such as startup. They are distinguished from the normal controls in that the critical variable is either normally uncontrolled or normally remains at a safe level as a result of the influence of other manipulated variables. Emergency controls typically come into service when it is no longer possible to maintain the critical variable within safe limits through the normal channels.

A case in point is pressure control of a distillation column. Pressure is normally controlled by adjusting the rate of heat transfer from the condenser. Under conditions of extreme heat load or insufficient cooling, the condenser valve will be driven to its limit without being able to hold pressure at the normal setpoint. An emergency pressure controller should then be available, having a higher setpoint and connected to a different manipulated variable. It could override the steam valve through a low selector, as was done by the high differential-pressure controller in Fig. 9-4. Or it could operate an auxiliary cooling valve or vent valve.

The emergency controller must be protected against windup so that it will not allow the critical variable to overshoot. If it acts through a selector, feedback taken from the selector output is all that is necessary, as in Fig. 9-4. If the emergency loop is independent of others, the most reliable approach is to use proportional or proportional-plus-derivative control. For certain level, pressure, and temperature loops this will be satisfactory, in that a narrow proportional band can be used stably.

For other variables, such as flow, integral action is mandatory, and the controller then must be equipped with an antiwindup circuit. The most notable example of this type is the "antisurge" controller for compressors shown in Fig. 9-7. A minimum flow must be maintained to avoid operating in the region where the compressor exhibits internal instability, known as "surge." Flow normally exceeds this value so that the antisurge flow controller keeps its recycle valve closed. Should the load on the compressor fall to the minimum allowable flow, the controller must open its valve before the setpoint is crossed; but since the recycle valve represents an artificial load, it should not be opened unnecessarily.

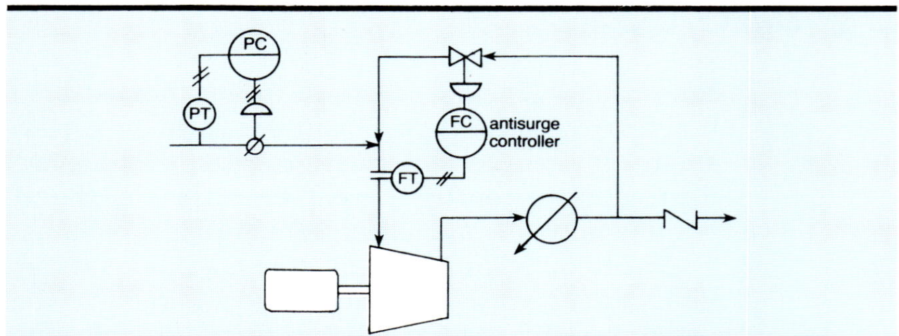

Fig. 9-7. The antisurge controller opens its valve only when the measured flow falls to its setpoint.

The antisurge controller also is used for startup purposes. Compressors are typically started into a closed system and must build enough discharge pressure to overcome the pressure in the pipeline before flow can begin. To avoid surge under these conditions, the antisurge controller opens the recycle valve before the compressor is started, closing it when a normal flow pattern is developed.

9-6. Operator Intervention

The purpose of constraint controls is to keep the plant operating in a safe manner under a variety of conditions—without manual intervention. It is important, however, for the operator to be made aware of the current mode of the plant, particularly if the mode is considered abnormal. This always will be the case when a product-quality controller has been overridden to favor some constraint. The mechanism for providing this information is a deviation alarm which is activated when the manipulated variable no longer follows the output of the normal controller. In the case of an emergency controller which manipulates its own valve, an alarm should be activated when that valve is leaving its normal (e.g., closed) position.

When multiple controllers manipulate a common valve through a selector, it is possible to provide operator access with a single manual station (HIC) as shown in Fig. 9-8. However, this compromises the integrity of the system: A failure in any input requiring operator intervention will cause the loss of all control. If each controller has its own auto-manual transfer capability, he can transfer the affected controller alone to manual, retaining the protection automatically provided by the others.

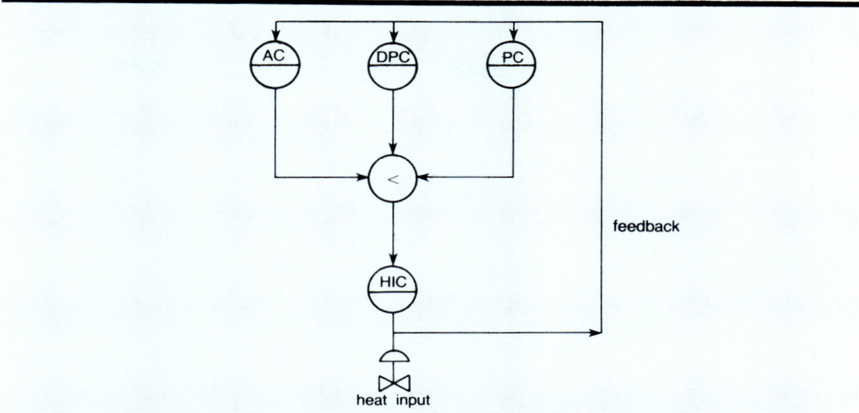

Fig. 9-8. The use of a single transfer station (HIC) deactivates the constraint controllers whenever the normal controller cannot be operated in automatic.

Where manual transfer at the valve is essential, additional feedback loops can protect against an unsafe condition, even in case of an operator error. Figure 9-9 shows a combustion control system for a boiler wherein fuel and air flow measurements are fed back to each other's setpoints to prevent the accumulation of excess fuel. The heat input demand sets both fuel and air. But if air flow fails to respond to an increasing setpoint, its flow is selected to set fuel flow. Similarly, if fuel flow exceeds heat demand for any reason, its value is selected to set air flow.

This sytem protects against a failure in either of the flow channels (but not both). Loss of a fan that reduces air flow will automatically reduce fuel flow in proportion. Similarly, the operator may manipulate either fuel or air (but not both) manually without losing protection.

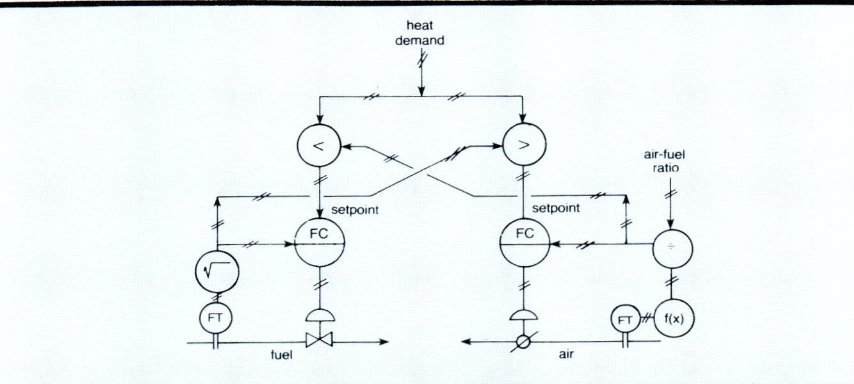

Fig. 9-9. The cross-limiting feature of this parallel fuel-air system prevents excess fuel from accumulating, even if one of the flow controllers is constrained or in manual.

To achieve this cross-limiting requires that both flow channels have the same setpoint. Fuel-air ratio correction then must be applied on the measurement side of a controller, as shown in Fig. 9-9. Increasing the air-fuel ratio applied to the denominator of the divider reduces the signal to the controller, causing it to increase air flow.

Exercises

9-1. *The high limit of a pneumatic controller can be set by its air supply. Does this prevent "windup" above the limit? How?*

9-2. *Add a column pressure controller and a base liquid-level controller to the system of Fig. 9-4. Also provide an alarm for override of temperature control.*

9-3. *Could antisurge control be achieved by overriding the suction valve in Fig. 9-7, instead of using a recycle valve? If not, why not? If so, what are the disadvantages?*

9-4. *Oxygen flow to an ethylene-oxidation reactor is manipulated to control oxygen content in the reactor below the lower explosive limit. Design a control system that uses redundant oxygen analyzers. Add a high-temperature shutdown.*

9-5. *The system of Fig. 9-9 is to be used in a fluid-bed boiler which has a high bed-temperature limit. Add a bed-temperature controller.*

Unit 10:
Variable System Structures

UNIT 10

Variable System Structures

This unit describes control systems whose structures can change with plant operating conditions.

Learning Objectives — When you have completed this unit, you should:

A. **Recognize the need to change system structure with changing plant conditions.**

B. **Be able to design systems that change from one structure to another smoothly.**

C. **Understand how to maintain loop gain constant as system structure changes.**

10-1. Unacceptable Loss of Control

The previous unit addressed the problem of encountering constraints—how to maintain control at constraints and avoid windup in those controllers which are not controlling. When a constraint was encountered, either in the form of a limited manipulated variable or in the form of a constrained controlled variable, control was yielded.

In the multiple-controller system illustrated in Fig. 9-4, the normal controller (temperature) was forced to yield to a constraint controller to avoid exceeding safe limits of operation. Under normal conditions, differential pressure floats between the constraints, and neither DPC needs to exercise control. They each, then have a "safe" condition where control is not required. The same is true of the temperature controller. At high-production conditions, when the high DPC reduces steam flow, temperature control is lost, allowing off-specification product to be made. But under low-production conditions, when the low DPC raises steam flow, loss of temperature control results in product that exceeds specifications. There is then no penalty, except the unavoidable expenditure of energy to keep differential pressure at its low limit.

If temperature is allowed to exceed its setpoint (high purity), but not fall below it (low purity), then an alternative manipulated variable must be provided when steam flow is constrained by high differential pressure. The principal difficulty encountered in arriving at an alternative system structure is finding an alternative manipulated variable. If one can be found, it cannot be expected to control as well as the primary manipulated variable, or it would have been chosen as the primary. This is not universally true, however—the choice of manipulated variable may be a function of plant operating modes that are substantially different, rather than different only in degree. Then a clear choice may be indicated.

A case in point is the control of pressure within a distillation column. If noncondensible gas is present, a vent valve must be manipulated; but if the vapor is totally condensible, there will be no venting, and the rate of heat removal or addition must be manipulated. Pressure must be controlled in either case.

The above two examples illustrated the need to change structures based on changes in the plant load. Another reason to change control-loop assignment is a fundamental shift in the quality of the manipulated variable. This is seen in the control of room environment—outside air may be used for cooling in winter, but for heating in summer. This manipulated variable then needs to be reassigned from one controller to another, seasonally. Ideally, this reassignment should be automatic, based on actual conditions rather than simply on time of year.

10-2. Duplication of Controllers

An issue which must be addressed is whether to use two controllers to cope with two distinct operating modes. Using two controllers keeps the system structure as simple as possible in that the two loops require no interconnection. Furthermore, each is adjusted for best response with its own manipulated variable. Consider, for example, the selector system of Fig. 9-8. Another pressure controller (not shown) normally manipulated condenser cooling. When the condenser can no longer remove all the heat put into the column, the cooling valve will open wide, and column pressure will rise. The PC in Fig. 9-8, having a higher setpoint, will begin to reduce heat input when its setpoint is reached. But, because there is no direct connection between the two manipulated variables, pressure is uncontrolled between the two setpoints—the condenser setpoint acts as a low-pressure limit, and the heat-input override acts as a high-pressure limit. If separate setpoints are undesirable, a single controller must be used, with explicit transfer of manipulated variables as described in Sec. 10-3.

In some applications, a neutral zone where the controlled variable is allowed to float may be advantageous. This is particularly true where control with *either* manipulated variable is costly and this cost can be saved. Consider a steam header whose pressure can be raised by letting steam down from a higher-pressure header, or lowered by venting steam. The former is costly because it loses work that could be recovered in a turbine, while the latter loses both heat and conditioned feedwater. A single controller could manipulate both valves in sequence, but one always would be open; then, a cycling controller or cyclic disturbance would be letting down steam only to vent it. Separate controllers would allow steam pressure to swing between their setpoints, and small fluctuations between supply and demand could be absorbed by the capacity of the header. Increasing the width of this dead zone saves energy, although at the cost of upsets to the users of the steam.

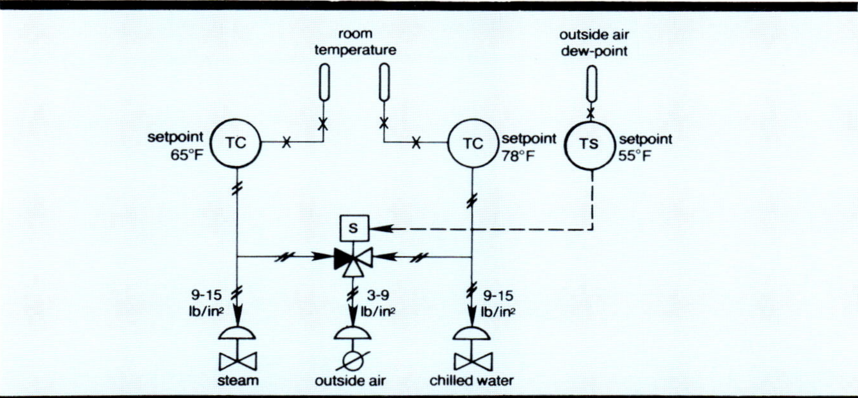

Fig. 10-1. Room temperature is allowed to float between the two setpoints; outside air is used for cooling in winter and heating in summer, as selected by its dew-point temperature.

The use of a dead zone is now quite common in environmental controls, where neither heating nor cooling is provided between, for example, 65 and 78°F. Figure 10-1 shows two temperature controllers establishing these limits. An outside-air damper is automatically transferred from one loop to the other as a function of ambient dew-point temperature. When ambient dew-point temperature is below 55°F, outside air is used to keep the temperature below 78°F; when the dew point exceeds 55°F, outside air is manipulated to keep room temperature from falling below 65°F. The latter is not a likely occurrence, but the former is, since people, machinery, and solar radiation all generate heat. Note that the outside-air damper is sequenced with the steam and chilled-water valves, so that the free form of heating and cooling always is used before the costly one.

10-3. Transfer of Manipulated Variables

Return to the case where pressure control must be transferred from condenser heat removal to a vent valve. If heat removal is modulated by the flow of cooling water, then a fully open water valve indicated loss of control. A single pressure controller can then manipulate a water valve and a vent valve in sequence, if their valve positioners are appropriately calibrated.

A more common means of adjusting heat removal is by flooding the condenser with condensate. This reduces the surface area available for condensation and thereby restricts heat transfer. However, maximum cooling is not indicated by a fully open valve, but by loss of liquid level within the condenser. It is imperative that liquid level not be lost altogether, or the reflux pump may be damaged. Therefore a liquid-level controller must override the pressure controller at this point, as shown in Fig. 10-2.

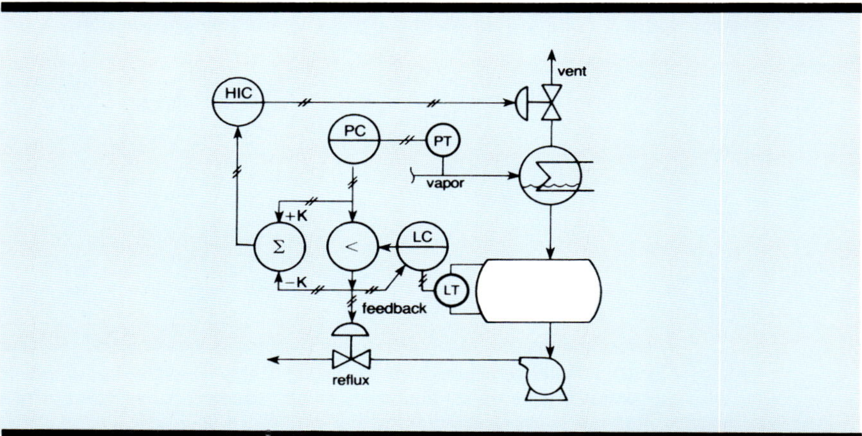

Fig. 10-2. Pressure control is transferred to the vent valve when the level controller is selected to manipulate the reflux valve.

The outputs of the pressure and level controllers are compared and the lower is selected for reflux manipulation. As long as the pressure controller is selected, the two inputs to the subtractor are equal, so that their difference—zero—keeps the vent valve closed. But when the level controller is selected, the subtractor develops a positive output, opening the vent valve. Pressure control is thereby smoothly transferred to the vent valve.

The pressure controller should be adjusted for stable performance when manipulating reflux. Since loop gain may differ when the vent valve is manipulated, the subtractor needs a gain adjustment common to both input signals. Then loop gain may be adjusted independently for both modes of operation. If dynamic responsiveness also differs significantly (it is not expected in this example), a lead-lag compensator can be added between the subtractor and the HIC. Note that the level controller needs external feedback to prevent windup, but the pressure controller does not because it is always controlling.

Occasionally, manipulation needs to be transferred to a valve that is not necessarily closed, as was the vent valve, or to a secondary controller whose setpoint is not zero. This occurs when a single pressure controller must override heat input, being no longer able to manipulate heat removal. Figure 10-3 shows the same level control override of reflux as Fig. 10-2, but now the subtractor biases the heat input downward to control pressure. If the pressure and temperature controllers are allowed to compete freely for the heat input in this mode, the pressure controller will drive the valve downward and the temperature controller will drive it upward, neither being satisfied. Competition in Fig. 10-3 is prevented by feedback of the valve position to the temperature controller externally. Then in the override mode, the feedback signal falls below the output of the temperature controller, avoiding windup.

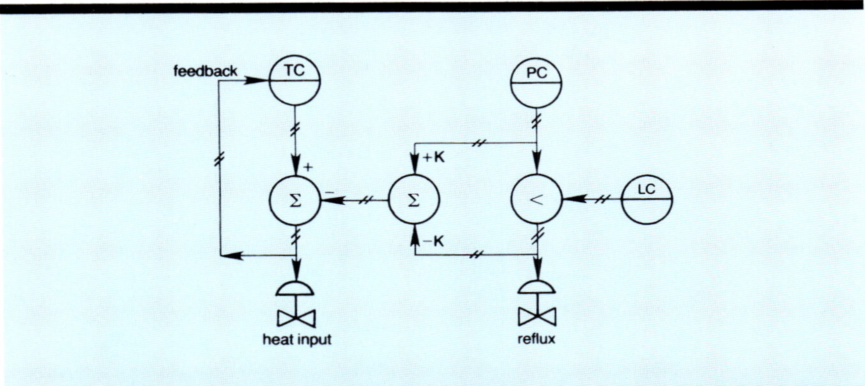

Fig. 10-3. Level override of reflux causes the pressure controller to reduce heat input; the external feedback loop stops the temperature controller from integrating in the override mode.

10-4. Multiple Similar Outputs

A very common problem is the manipulation of more than one final element to affect a single controlled variable. The final elements could be identical, similar in nature but different in size, or opposite in effect. In its simplest form, this involves sequencing a pair of valves so that one always opens before the other; or when they have opposite effects, as in heating and cooling, they are sequenced so that one closes before the other opens.

In many situations, however, it is desirable to alter the sequence manually, or through automatic overrides, to protect equipment, equalize duty, or optimize performance. This is particularly true when many manipulated variables are arranged in parallel service.

To illustrate the point, consider the control of air flow to a boiler by manipulating dampers on parallel fans, as in Fig. 10-4. Here control signals are shown to be electronic because instantaneous transmission is essential for this system to function. The output of the flow controller enters as a positive input to a high-gain amplifier. Feedback from the HIC stations enters as negative inputs scaled so that their sum equals full-scale controller output. (If the capacities of the two dampers are equal, each signal is given half the gain of the controller output, as shown.)

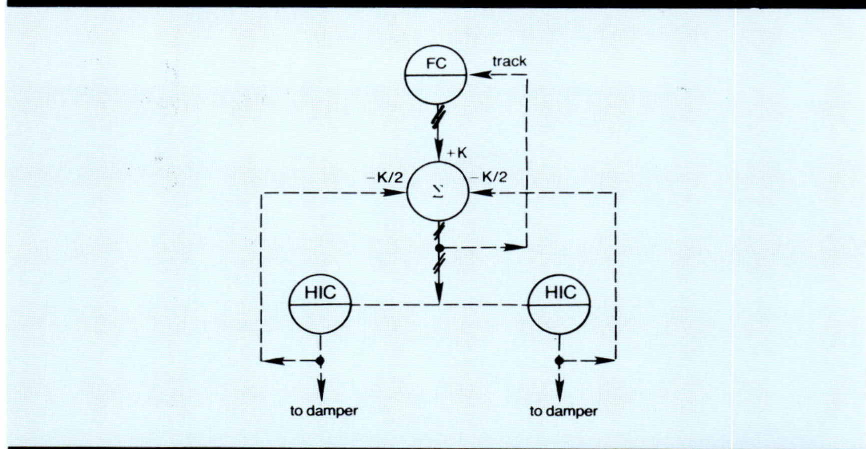

Fig. 10-4. A high-gain summing amplifier drives the dampers so that their average position matches the flow-controller output.

If both HIC stations are in automatic and neither output is limited, a change in the FC output will be passed equally to both dampers, and both will respond equally. But if one HIC is placed in manual, its damper will not respond to the output of the high-gain amplifier, thus reducing the feedback by half. This causes its forward-loop gain to double, so that the damper remaining in automatic is driven twice as hard as before. In this way, the gain of the flow-control loop is the same whether one or two dampers respond to its output.

Operator-induced upsets also are eliminated by this system. When the operator moves one damper in manual, its feedback signal causes the amplifier to drive the other damper an equal amount in the opposite direction; in effect, the average of the two feedback signals always is driven to match the controller output. The HIC stations also may be equipped with limiters and bias adjustments, allowing the operator to balance fan loading at his discretion. None of these actions can disturb flow control or change its loop gain, as long as one of the dampers is free to move in automatic.

An additional provision often is included to cover the situation when *all* HIC stations are in manual, as during startup. A logic circuit places the FC in a "track" mode when all HIC stations are in manual. In this mode, the FC is driven by the summing amplifier until its output matches the average of all the damper positions —in essence the normal functioning of the system has been reversed. When the first HIC is then placed in automatic the summing amplifier is already balanced, and there will be no bump.

In pneumatic systems, and in systems where feedback is taken from flowmeters, a high-gain amplifier cannot be used, owing to the delay in signal transmission. It can be replaced by an integrator or total-flow controller, as shown in Fig. 10-5. In this example, two fuel flows are manipulated in parallel to satisfy a common heat demand. The flow controller must be adjusted for stable performance when all loops are in automatic because that is when its loop gain is highest. In effect, it takes the burden of variable loop gain away from the pressure controller; this practice is acceptable since the flow loop is so much faster than the pressure loop.

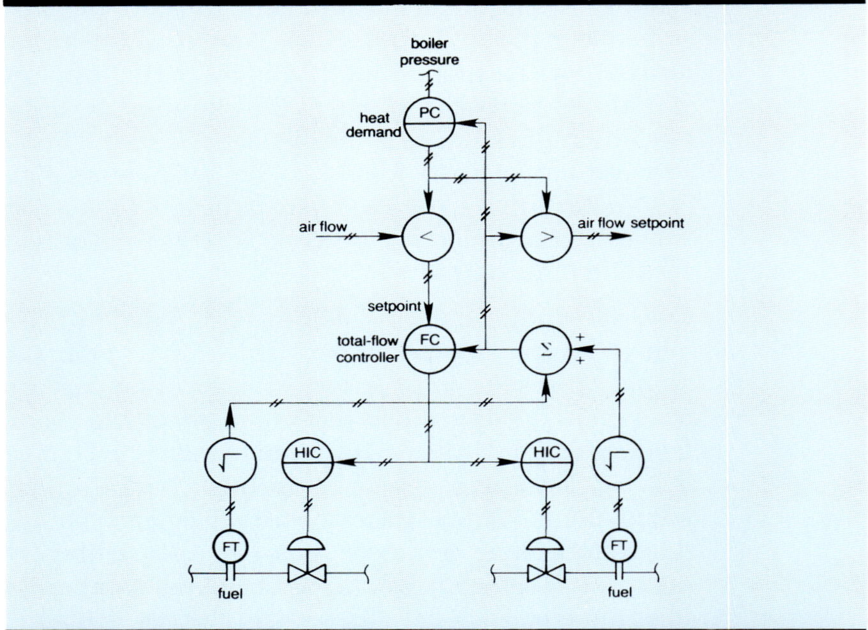

Fig. 10-5. The total-flow controller attempts to satisfy heat demand regardless of which fuel is available for manipulation.

The management of the fuels takes place below the level of the total-flow controller, at the HIC stations. The fuels may be ratioed to each other, sequenced, limited, or placed under overrides from gas pressure or tank level. While all these actions affect individual flow rates or their relationships with each other, total flow still will be maintained to satisfy the demand for heat. The summing amplifier here must be calibrated on a heat-input basis, since that is demanded by the pressure controller and by the air flow system. Observe that total heat input also is sent to the PC as external feedback. When all HIC stations are in manual,or signals are otherwise limited, the FC can windup but the PC cannot.

A refinement of this approach has individual flow controllers for each fuel so that flow setpoints may be adjusted relative to each other rather than valve positions. This allows more accurate management of the fuels, but introduces another level of cascading. The total-flow controller then cannot be adjusted for as tight control as before, owing to its interaction with the individual flow controllers, which respond to the same measurements. For best results, the individual flow controllers should be adjusted for tight proportional action, while the total-flow controller is tuned for dominant integral action. This will move the periods of the loops apart and reduce their interaction.

10-5. Multiple Dissimilar Outputs

Here we consider the case of manipulating a group of on-off devices in parallel with a single modulating device. The modulating device is necessary to balance a variable process load and therefore attain a steady state—but only one is needed. On-off devices can be brought on or off line when the range of the modulating device is exceeded.

An example of this application is the delivery of compressed air to a common header by a group of constant-speed compressors. To match supply exactly with demand, and thereby control header pressure at a steady state, it is necessary to have one variable-speed driver, or a throttling valve. Because the variable-speed compressor or valve can accommodate only a fraction of the full load, the constant-speed compressors need to be brought in and out of service as major load changes occur. But this must be done smoothly if tight pressure control is to be achieved.

Figure 10-6 shows how this might be accomplished. The output of the pressure controller represents the full range of supply from all compressors. Pressure switches are set to start compressors at increments of demand approximately equal to their individual capacities. The flow developed by the constant-speed units is subtracted from the total demand to produce the remaining demand, which the modulating unit must supply. The gain K of the subtractor is the ratio of total supply capacity to the capacity of the variable unit.

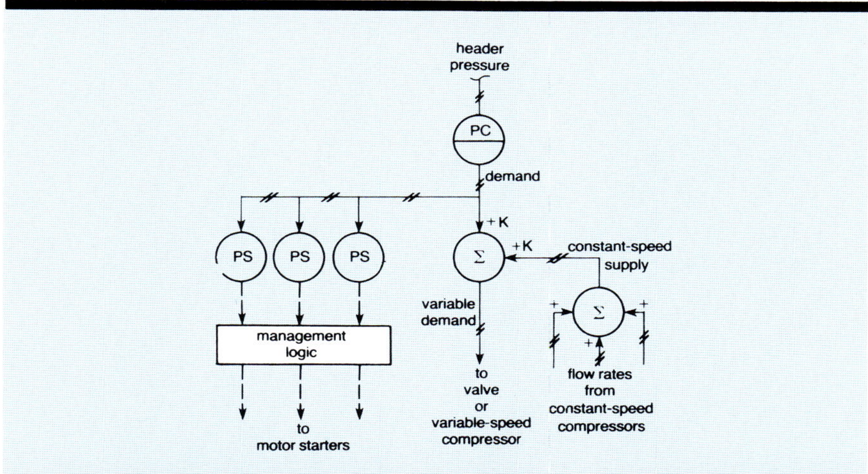

Fig. 10-6. The bulk of the demand is allocated to constant-speed compressors by the pressure switches; the balance is sent to the valve or variable-speed compressor.

This system allows for manual intervention without upsetting the controlled variable. Because actual flow is fed back, the modulating unit will respond directly to changes in the other units. The only restriction is its limited range.

It is assumed that the on-off compressors have about equal capacity. This allows them to be rotated in and out of service either automatically or manually, to equalize wear. If they are not equal, management becomes more complex. A binary series of compressor sizes will cover a wide range of loads with fewer machines, but it requires dedicated service. An analog-to-binary converter would replace the pressure switches shown in Fig. 10-6.

10-6. Cascaded Multiple Outputs

In an industrial boiler plant, there may be several boilers supplying a common header, and each boiler may have more than one supply of fuel. To provide complete integration of all fuels which affect the single steam pressure, multiple-output systems must be cascaded.

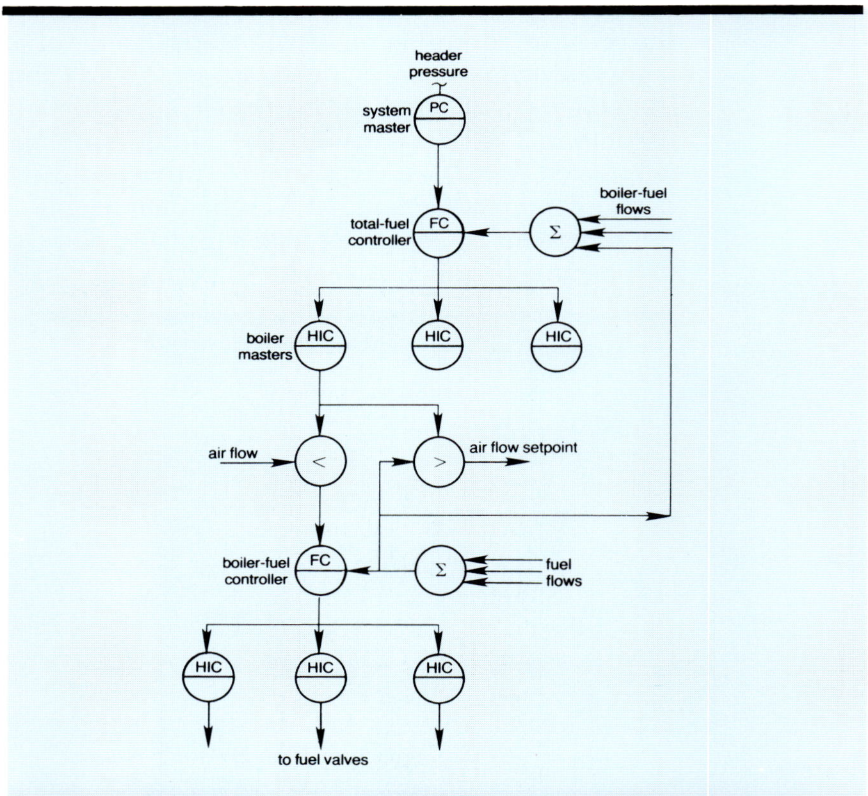

Fig. 10-7. While header-pressure control requires only total-fuel manipulation, each boiler needs fuel and air controls for safe operation and load allocation.

In Fig. 10-7, the master pressure controller sets total fuel flow to all boilers. The total-fuel FC then places demands on individual boiler fuel-flow controllers. While this creates a flow-on-flow cascade, it is unavoidable, in that each boiler must have its own fuel-control system for coordination with air flow in a safe and efficient manner.

Exercises

10-1. *Identify the variably structured systems described in Unit 2 and comment on their use.*

10-2. *Repeat for Unit 3.*

10-3. *Two fuels are used to fire a boiler. One is a waste gas, of limited availability; the other is natural gas. Devise a system that uses as much waste gas as is available, then supplement with natural gas as necessary. The steam pressure-control loop should function equally well on either fuel, or both together.*

10-4. *Consider a two-fuel system such as above, where the waste fuel is wood. If burned in excessive amounts, or if it is especially wet, smoke is formed. Arrange a system to limit wood firing based on stack opacity.*

10-5. *A chilled-water refrigeration system consists of eight parallel constant-speed compressors, each with its own bypass valve for controlling its water discharge temperature. What is the most efficient means for distributing refrigeration load? Which of the systems described in this unit is most applicable?*

Unit 11:
Inferential Control

UNIT 11

Inferential Control

This unit addresses the control of processes whose important variables are unmeasurable and do not yield to direct calculation; models of the process then must be used to correlate measurable variables in such a way as to provide regulation of those which are unmeasurable.

Learning Objectives — When you have completed this unit you should:

A. Recognize those processes requiring inferential control.

B. Understand how measurable variables are related in the process model.

C. Recognize the presence of feedback loops and provide compensation for them.

11-1. Controlled Variables

Certain processes exist which require close control over variables which cannot be measured directly. In many cases, the variable requiring control is a product composition which can be determined only by laboratory analysis. Direct feedback control cannot be employed, therefore some substitute must be used. A substitute measurable variable may be available for feedback control, which would only require periodic setpoint adjustments based on analytical results. If no single measurement is available, some combination of two or more may suffice. As the difference between the true controlled variable and the measurable variables increases, their inferential relationship becomes more complex and less accurate.

This concept is illustrated by antisurge control of a centrifugal compressor. As introduced at the end of Unit 9, antisurge control requires maintaining flow above some minimum value to keep the compressor out of an unstable operating region. Identifying the point at which this region is entered is difficult, however, in that it depends on other factors besides flow.

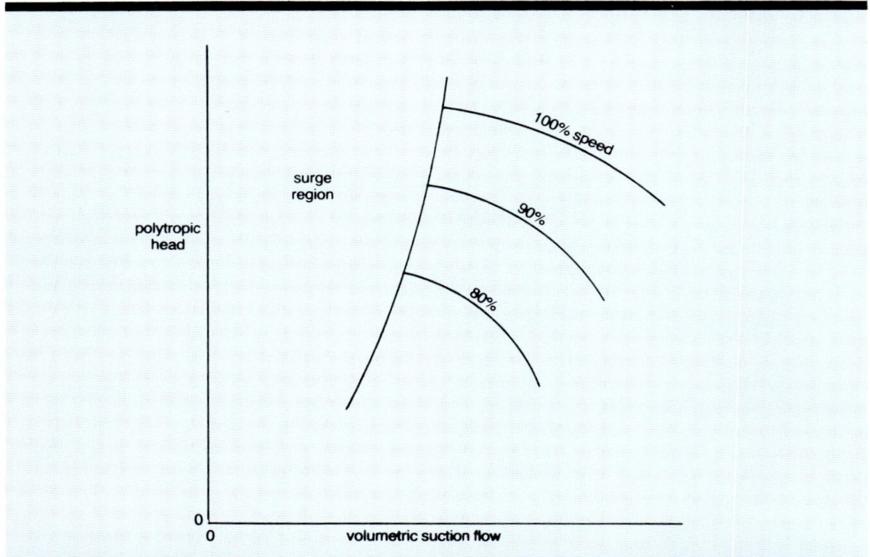

Fig. 11-1. Surge develops at a critical juncture of head, flow, and compressor speed.

A set of characteristic curves for a typical centrifugal compressor appears in Fig. 11-1. The unstable "surge" region is identified. Observe that the coordinates are polytropic head and volumetric suction flow, neither of which is measurable. If both were directly measurable, an equation could be fitted to the surge line, and its solution could be the input signal to the antisurge controller.

The measurements which are available include suction and discharge pressure and temperature, suction or discharge orifice differential pressure, rotational speed or inlet guide-vane position. Speed is the only measurable variable appearing in Fig. 11-1. Polytropic head includes compression ratio and suction temperature, along with the specific-heat ratio, supercompressibility factor, and molecular weight of the gas. The volumetric suction flow is a function of orifice differential pressure, temperature, absolute pressure, and gas molecular weight.

While it is possible to calculate the exact relationship between measurable variables defined by the surge curve, as described in Ref. 1, this does not completely solve the problem. The gas composition could change, and the surge curve described by the compressor manufacturer does not necessarily fit the machine exactly under operating conditions. Therefore a preferred strategy is to develop a simple model which can be adjusted easily by the operator.

This approach is followed in Fig. 11-2. Several points taken from the manufacturer's curve are converted into equivalent values of compressor differential pressure and orifice differential pressure, using the relationships given in Ref. 1.

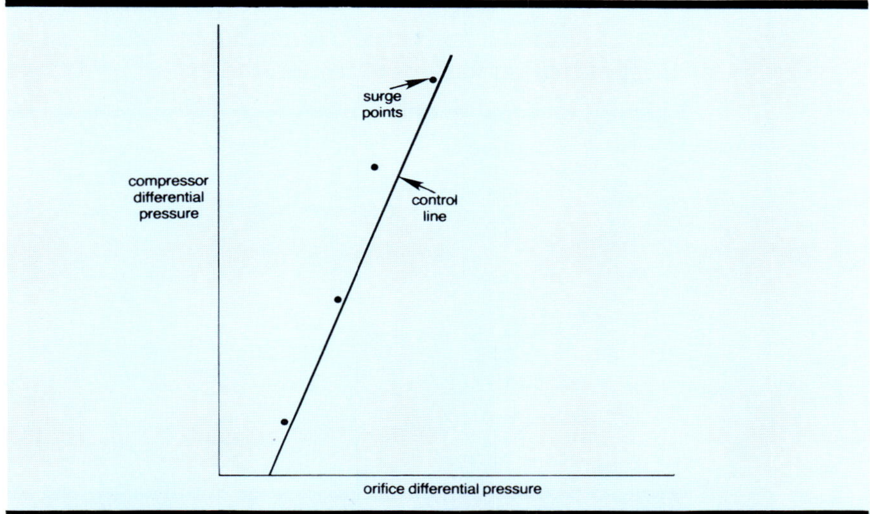

Fig. 11-2. A simple relationship between easily measured variables gives the most operable system.

They fall in nearly a straight line, allowing a linear model whose slope and intercept are adjustable by the operator. Then the line may be repositioned if the compressor should begin to surge without control action being taken, or if the recycle valve should be opened unnecessarily. The structure of the antisurge system is given in Fig. 11-3. This is not a unique or universal solution to the problem—other combinations of measurements can be used, depending on their availability.

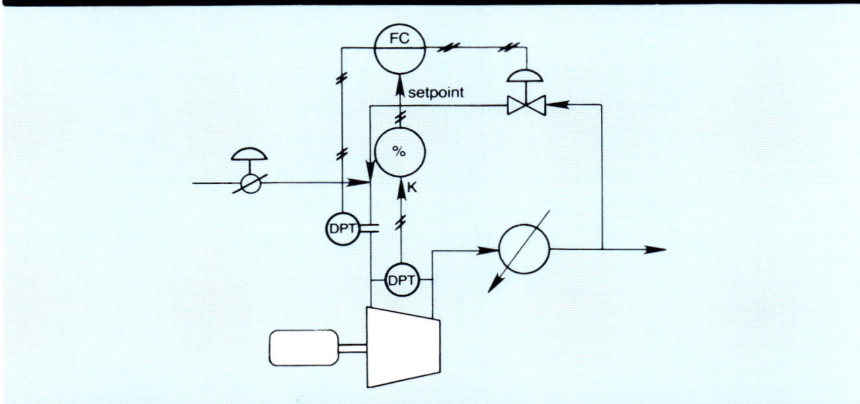

Fig. 11-3. The antisurge system uses compressor differential pressure to set the minimum flow.

11-2. Relating Manipulated and Controlled Variables

The antisurge system compared two controlled variables, compressor and orifice differential pressures, both of which responded to the manipulated recycle valve. In some other inferential systems, it will be necessary to relate a manipulated and controlled variable through the use of a process model.

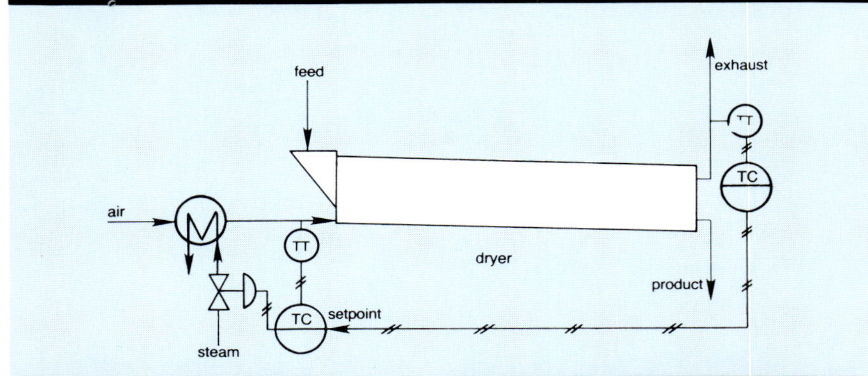

Fig. 11-4. Controlling exit-air temperature at a constant setpoint allows product moisture to change with evaporative load.

Figure 11-4 pictures a typical dryer wherein heated air encounters wet feed in cocurrent flow. The air cools as sensible heat is consumed in evaporating moisture. Cool air exits at the product end of the dryer. Product moisture cannot often be measured on-line, and therefore cannot be directly controlled. Instead, outlet-air temperature is customarily controlled, but this allows product moisture to vary with evaporative load.

Should the feed rate or moisture rise, evaporation increases, causing the outlet-air temperature to fall. The controller returns measurement to setpoint by raising heat input, but this will not completely accommodate the increased load. While the temperature difference between the air and the solid product increases at the inlet of the dryer, it actually decreases at the outlet, owing to a rise in solids temperature with heat input. The lower temperature difference at the product end causes product moisture to rise.

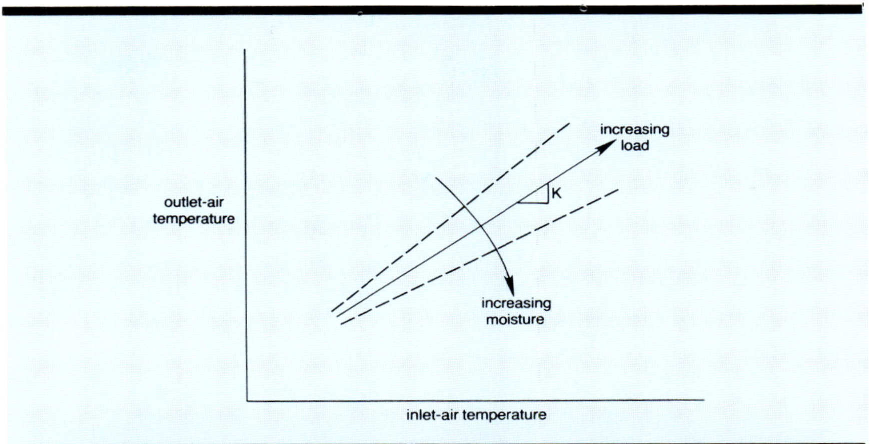

Fig. 11-5. The moisture in a product leaving a dryer can be regulated if outlet- and inlet-air temperatures are related as shown.

Using the models developed in Ref. 2, a relationship was derived which requires outlet-air temperature to rise with inlet-air temperature as shown in Fig. 11-5. This relationship is between a controlled variable (outlet) and a manipulated variable (inlet). Observe that changes in load affect outlet temperature alone. By comparison, changes in load affected both the compressor and orifice differential pressures in the antisurge system.

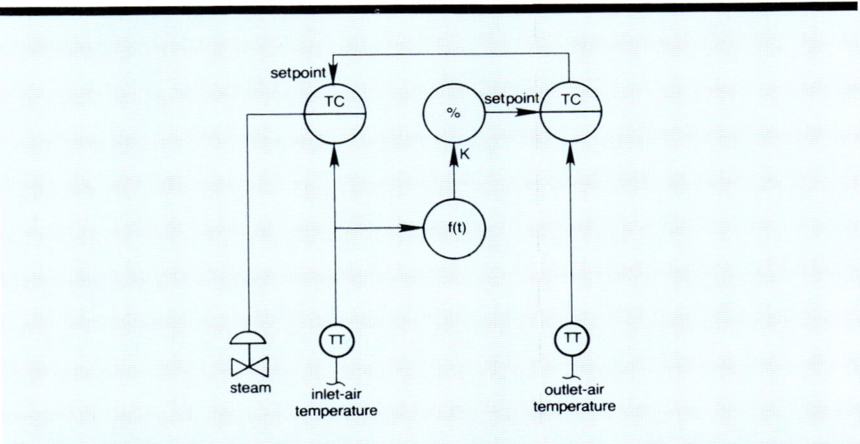

Fig. 11-6. The outlet-air TC sets the inlet-air TC in cascade; then the outlet-air temperature setpoint is programmed as a function of the inlet-air temperature measurement.

The inferential control system appears in Fig. 11-6. On an increase in evaporative load, outlet-air temperature begins to fall, as before, and the controller responds by raising the inlet-air temperature setpoint. The system then calculates a new, higher setpoint for outlet-air temperature to satisfy the relationship required for constant product moisture. This reaction causes heat input to rise again, increasing both temperatures further, along the path described by Fig. 11-5. Eventually, the system will come to rest with both temperatures higher, and product moisture still controlled.

There are factors other than temperature which affect product moisture, however. Atmospheric humidity can be significant, particularly at low drying temperatures. Additionally, the slope of the curve in Fig. 11-5 is not the same when feed moisture varies as when feed rate varies, except for fluid-bed dryers. The relationship is further affected by particle-size distribution. Yet the system shown in Fig. 11-6 was kept simple to make it easy to operate and adjust.

The relationship between the two temperatures is adjustable both in slope and intercept, to produce material of the desired moisture content both at heavy and light loads, and under a variety of atmospheric conditions. Adjustments are made after comparing laboratory analyses of product moisture with desired values. To reduce product moisture requires an increase in slope.

The availability of an on-line analyzer does not necessarily eliminate the need for the inferential system. Most dryers are dominated by dead time in the transportation of solids. By the time off-specification product has reached the analyzer, there may be much accumulated in the dryer that cannot be upgraded by control action. Because the inferential system operates on rapidly responding air temperatures, load upsets are detected earlier than by a product analyzer, and control action can be applied in a timely, if imprecise, manner. The principal role of the analyzer then would be to update the inferential model. A composition controller could automatically adjust the slope K by means of a multiplier.

Reference 3 describes another type of dryer, where heat is applied directly to the product through contact with steam tubes. Temperatures in this dryer are not as sensitive to changes in load or heat input, but steam pressure in the tubes and steam flow are. An inferential control system has been designed using steam pressure as a controlled variable, whose setpoint is programmed as a function of steam flow.

11-3. Inferring Load Variables

Feedforward control can be very useful in reducing the sensitivity of controlled variables to load disturbances. It even may provide regulation in the absence of feedback in those cases where the controlled variable is unmeasurable. However, a true measurement of process load or demand is not always available and may have to be calculated or inferred from measurable variables.

Consider a boiler supplying steam either to a single user or to multiple users. Typically, the users place independent demands (which the boiler is expected to meet) on the steam header. Imbalance between supply and demand appears as a change in header pressure, which is then the primary controlled variable for operating the boiler. Due to delays in the firing circuit and heat capacity within the boiler, a changing steam demand may cause a large deviation in steam pressure to develop before a matching supply is generated. The delay can be reduced by setting firing rate proportional to demand, before a pressure deviation develops —this is the essence of feedforward control.

A steam flowmeter is usually available at the boiler exit. However, it indicates neither supply nor demand, but something between. The true demand for steam is the position of the user valves—this would be indicated as steam flow only if header pressure were absolutely constant. Since it is not constant, the position of the user valve or aggregate valves must be calculated.

The mass steam flow developed through a valve of opening C_v is approximately

$$W_s = C_v p \tag{11-1}$$

where p is the steam pressure at the header. The steam flowmeter develops a differential pressure h proportional to steam flow:

$$W_s = k \sqrt{hp} \tag{11-2}$$

These two equations may be set equal and solved for unmeasurable C_v in terms of measurable h and p:

$$C_v = k \frac{\sqrt{hp}}{p} = k \sqrt{h/p} \tag{11-3}$$

In this way, the steam flow measurement is not used alone for feedforward control, but is compensated for changes in pressure. Observe that pressure is in the denominator rather than in the numerator, where it appears in the mass-flow equation (11-2).

On opening a valve to draw steam, boiler pressure will fall, and steam flow will increase. This increment in steam flow is not supplied by an increment of generation, but is borrowed from the steam stored in the boiler. At this point, there is no increase in supply, only an increase in demand. In Eq. (11-3), numerator h rises and denominator p falls, both of which signal an increase in demand. If Eq. (11-2) were used, changes in h and p would tend to offset each other, thereby failing to reflect the true demand.

Boilers also can be upset on the supply side by unintentional changes in fuel and air flows and by variations in fuel quality. An increase in supply will raise steam flow and pressure at the same time, into a constant demand. The simultaneous change in numerator and denominator of Eq. (11-3) will produce a constant output signaling a constant load. The pressure controller must then readjust firing rate to counter the supply-side upset. This system is described in Fig. 11-7.

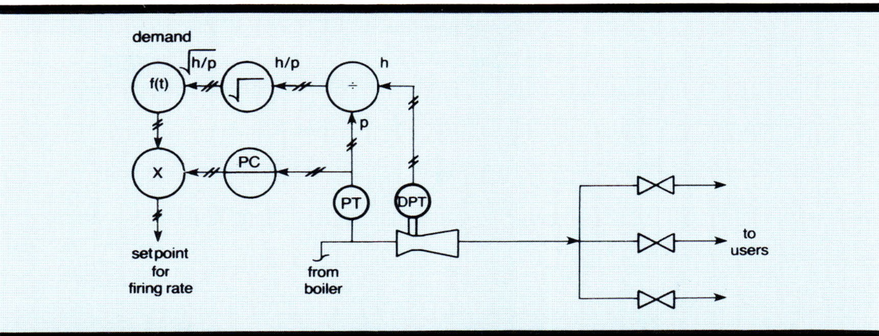

Fig. 11-7. Pressure compensation of steam flow using a divider is necessary to distinguish demand from supply.

11-4. Multiple Feedback Loops

Inferential systems make use of several measured variables to provide the information that is normally available in a single controlled or load variable. In the case of a controlled variable, a feedback loop is anticipated and the action of the controller is selected to insure that the feedback is negative. In the case of a load variable, no feedback is expected, so there is usually no consideration given to stability.

With the use of more than one variable, however, other feedback loops may be formed, which may have a pronounced effect on the stability of the system. In the case of the compressor antisurge system of Fig. 11-3, opening the recycle valve raises suction flow—the control action can be selected for negative feedback here. But compressor differential pressure also changes when the valve is opened—it decreases as flow is increased. Because compressor differential pressure is applied to the setpoint of the antisurge flow controller, it produces action opposite to that of the measurement input. Therefore both feedback loops in the antisurge control circuit are negative. The controller can be adjusted for stable performance if its gain favors their combined gains and its integral time favors the slower loop. Since both measurements respond at the same time in this process, no compromise is necessary, and controller adjustment is no more difficult than with a single loop, as long as it is performed in the remote-set mode, i.e., with the setpoint loop closed.

In the dryer control system of Fig. 11-6, a negative-feedback loop exists from the outlet-air TC through the inlet-air TC, the steam valve, and back to the outlet-air measurement. However, a second feedback loop appears—from the outlet-air TC to the inlet-air temperature measurement back to the outlet-air setpoint. This is positive in sense, paralleling the negative-feedback loop through the dryer, but being applied to the setpoint rather than the measurement of the outlet-air TC.

For a system containing a positive-feedback loop to be stable, its gain must be lower than that of the coexisting negative-feedback loop at all frequencies. If its steady-state gain is higher, the system will tend to wind up, coming to rest only at a limit.

If its steady-state gain is lower, but dynamic gain is higher, the positive-feedback loop will cause oscillations to develop. Positive feedback is the mechanism used to provide integral action in many types of controllers. In that role, its steady-state gain is set equal to the negative feedback internal to the controller; but its time constant also must be longer than the dynamic response of the negative-feedback loop through the process if stability is to be achieved.

In the case of the dryer control system, the gain K of the positive-feedback loop is set to provide regulation of product moisture, as required by the process model. For dryers examined in Ref. 2, the slope of the outlet-inlet relationship is in the range of 0.15 to 0.25. However, recognize that the steady-state gain of the negative-feedback loop through the process also may be low. A one-for-one correspondence between inlet and outlet temperatures only appears when the dryer is empty. At real loads, much of the sensible heat of the air is absorbed in evaporation and therefore does not appear in outlet temperature. But in the author's experience, those dryers requiring a higher value of K also have a higher gain in their negative-feedback loop. Steady-state stability is therefore achieved if the inferential system is properly calibrated.

A positive-feedback loop also appears in the boiler control system of Fig. 11-7. An increase in steam flow caused by opening a valve will, through the feedforward loop, increase firing rate. But raising firing rate increases steam production, which will appear in the steam-flow signal later, and thereby raise firing rate again. This feedback loop has a unit steady-state gain, although its dynamic gain is reduced by the heat capacity of the boiler.

Including steam pressure in the feedforward calculation adds negative feedback to the system. Increased firing raises pressure along with flow; but being applied to the denominator of the divider, this signal causes a subsequent reduction in firing rate. In practice, the steam flow and pressure signals are in phase, so their simultaneous application to numerator and denominator effectively cancels the positive feedback with an equal quantity of negative feedback. The result is the net elimination of feedback, so that the divider output contains only feedforward information.

11-5. Dynamic Compensation

If the signal paths which are combined to form the inferential system have different response characteristics, dynamic compensation may be necessary. In the compressor antisurge system, compressor and orifice differential pressures responded at the same rate, so dynamic compensation was unnecessary. The same is true of the boiler control system. The dynamic compensator f(t) appearing in Fig. 11-7 is not for the purpose of correcting a phase difference between the feedback loops, because it is applied to both. Its function is to overcome the dynamic lag in the response of steam flow to firing rate, to minimize the deviation in steam pressure following load changes. It plays the same role as it did in the decoupling systems of Unit 8.

The dryer control system of Fig. 11-6 does require phase balancing between its positive- and negative-feedback loops. Negative feedback is delayed by the heat capacity of the dryer and the transportation of air through it. The positive-feedback loop has no delay at all, unless a dynamic compensator f(t) is added as shown. Without this compensator, the system will tend to oscillate, much as it would if the integral time of the outlet-air TC were too short. The simplest form of compensator, a first-order lag, is sufficient for stabilization. It ought to be set to a value at least as great as that required by the integral mode of the outlet-air TC.

Like dynamic compensators in feedforward systems, its time constant ideally will be that which minimizes the change in product moisture following a disturbance in load. For those dryers whose residence time for product is much greater than that for air, the dynamic compensator should be set for a longer time constant than required for feedback-loop stabilization. In the absence of on-line moisture measurement of the product, setting the compensator approximately equal to the expected residence time of solids in the dryer should give acceptable results.

Exercises

11-1. *It is necessary to control the mass of particles contained in a fluidized-bed boiler, although it cannot be measured directly. The resistance of the bed to air flow varies with mass. Devise an inferential system for controlling the mass; it must function under variable air flow.*

11-2. *Describe two inferential systems commonly used for distillation-column control. To which class do they belong?*

11-3. *It is desired to control the average temperature inside an exothermic reaction mass, where no measurement may be made. Devise an inferential system to do this, using inlet and outlet temperatures of the cooling water. Its inlet temperature is manipulated under constant flow.*

11-4. *An air compressor is subject to rapid load changes. Devise a feedforward system to change compressor speed as a function of demand for air.*

11-5. *Identify the inferential system in Unit 3 and comment on its feedback loops.*

References

[1]Shinskey, F. G. *Energy Conservation through Control*. New York: Academic Press, 1978. pp. 130-137.
[2]*Ibid*., pp. 215-225.
[3]*Ibid*., pp. 243-248.

Appendix A: Suggested Readings and Study Materials

APPENDIX A

Suggested Readings and Study Materials

Independent Learning Modules:

Murrill, P. W., *Fundamentals of Process Control Theory*, (Instrument Society of America, 1981).

Handbooks:

Liptak, B. G., *Instrument Engineers Handbook: Vol. II—Process Control*, (Chilton Book Co., 1970).

Textbooks:

Buckley, P. S., *Techniques of Process Control*, (Wiley, 1964).
Harriot, Peter, *Process Control*, (McGraw-Hill Book Company, 1964).
Shinskey, F. G., *Process Control Systems*, 2nd Ed. (McGraw-Hill Book Company, 1979).
Shinskey, F. G., *Distillation Control: for Productivity and Energy Conservation*, (McGraw-Hill Book Company, 1977).
Shinskey, F. G., *Energy Conservation through Control*, (Academic Press, 1978).

Technical Magazines and Journals:

Control Engineering, published by Technical Publishing Co., New York.
ISA Transactions, published by the Instrument Society of America.
Instrumentation Technology, published by the Instrument Society of America.

Appendix B: Solutions to All Exercises

APPENDIX B

Solutions to All Exercises

UNIT 1

1-1. Flow is production rate; composition is product quality.

1-2. The interaction may be severe—it depends on the relative sizes of the two streams. If they are of equal size, both will affect flow equally, maximizing interaction.

1-3. Every change in flow will upset composition.

1-4. One of the variables must be left uncontrolled—it is usually steam flow, which is subject to independent demand from users.

1-5. Yes, there are likely to be more measurements than manipulated variables. The extra measurements can enhance control through the use of cascade loops and feedforward control.

1-6. Unless the mathematical relationships among the process variables are known, and used to configure the control system, the system may not perform equally well under all operating conditions. Also, as the number of variables increases, the intuitive approach is more prone to error.

UNIT 2

2-1. There is intermediate storage between the crude fractionator and the downstream reactors. Reactor feed rates are adjusted manually from time to time as inventory in the storage tanks changes.

2-2. Unless steam supply is set equal to demand at all times, steam pressure will vary; the variation is a function of the rate-of-change of demand and the dynamic response of the boiler. The response can be improved through feedforward control, setting firing rate proportional to steam demand (see Sec. 11-3).

2-3. Ratios in Fig. 2-3 are set between individual ingredients and base stock, whereas in Fig. 2-4 they are set against total flow. The numbers will differ by the fraction of base stock in the blend.

2-4. Using a total-flow measurement to set the system in Fig. 2-4 would form a positive feedback loop with unity gain. It would not start or regulate—it would have no direction.

2-5. The VPC prevents loss of ratio control caused by excessive flow of base stock. An analyzer control loop would operate on the additives alone and so could not replace the function of the VPC.

2-6. A turbine trip would simultaneously close the turbine throttle and open the bypass. The deviation from zero setpoint for VPC-3 will tend to be greater in this condition than it was during startup, so that the downward ramp will tend to be faster than was the upward.

UNIT 3

3-1. Self-regulation allows a process to find a stable state even without automatic control—an advantage during times of multiple-loop failures, as when a computer control system goes out of service.

3-2. No. Small changes in level will have negligible effect on the pressure drop between feedwater pump and boiler, which determines feedwater flow.

3-3. Boiler steam pressure is self-regulating when steam demand is set by users with manual valves. Then changes in pressure will cause proportional changes in flow. If demand is set by mass-flow controllers, there will be no self-regulation.

3-4. The column base level controller admits bottom-product flow into the surge tank. The surge tank level controller then adjusts column feed rate with proportional action. The additional capacity of the surge tank above that of the column base would make this loop more stable than that in Fig. 2-2.

3-5. Liquid levels are controlled either at the high or low setpoints, but not between. Then an outage somewhere in the line that causes the direction of control to be transferred will also cause one of the levels to drift from one setpoint to the other.

3-6. Figs. 2-7 and 2-8 are bidirectional control schemes for a boiler supplying steam to a turbine.

UNIT 4

4-1. Discharge pressure falls as flow is increased, due to internal resistance.

4-2. Many reactions will not proceed until a temperature is reached that is well above ambient. Fires need to be lighted, engines need to be started.

4-3. Partial pressure can be controlled in a feedforward manner by setting the hydrogen-to-hydrocarbon ratio as a function of measured total pressure. Hydrogen partial pressure divided by total pressure is the mol fraction hydrogen in the feed:

$$\frac{p_H}{p} = \frac{H_2}{H_2 + HC}$$

where HC is the flow of hydrocarbon feed. Solving for H_2 setpoint:

$$H_2{}^* = \frac{HC}{\dfrac{p}{p_H{}^*} - 1}$$

where $p_H{}^*$ is the desired partial pressure.

4-4. If pressure is allowed to vary rapidly, this develops a temporary difference between actual temperature and boiling point. As a result, the rate of boiling may change drastically, causing flooding of the trays.

4-5. If compensation is chosen instead of control, a manipulated variable is eliminated along with its associated cost. However, the engineer must be certain that the environmental variable will not change so fast as to create problems for the process, as in 4-4 above.

4-6. The actual flow of heat to the reboiler can be calculated from oil flow and the temperature difference between inlet and outlet. This can be the input to a heat-flow controller which manipulates the oil valve. The controller will react to changing inlet temperature by adjusting flow accordingly.

4-7. Refrigerant temperatures or pressures at both hot and cold sides of the system may be allowed to float by eliminating restrictions to heat transfer.

UNIT 5

5-1.

$$\frac{Q_2}{Q_1} = \frac{2 \ln 0.02}{2 \ln 0.01} = 0.849$$

Approximately 15% savings.

5-2.

$$\frac{Q_2}{Q_1} = 0.5 \frac{\ln 0.02}{\ln 0.003} + 0.5 \frac{\ln 0.02}{\ln 0.017} = 0.816$$

Approximately 18.3% savings.

5-3. In the cycle appearing in Fig. 5-2, much more time is spent at the lower level of impurity than at the higher; therefore the estimate above may be conservative.

5-4. The sample should be taken from the overhead vapor line at the inlet to the condenser. This will eliminate time lags in the condenser and reflux drum, as well as the longer transportation time needed for liquid samples.

5-5. Column temperature can be set in cascade by a composition controller. Alternately, rate of temperature change can be added to the analysis for input to a single controller, as in Fig. 5-3. This eliminates a control station and simplifies tuning, but does not allow temperature control when the analyzer is out of service.

5-6. The programmed adaptive controller is always in proper adjustment based on current information on process conditions. The feedback adaptive system depends on process response to diagnose performance, and is therefore always lagging behind the true state of the process.

UNIT 6

6-1. a) Conversion, yield, catalyst life, production rate, purge rate.

b) Product recovery, condenser valve position, steam flow per unit product or feed, concentration of product in byproduct stream, total operating cost.

6-2. Production rate, purge rate, condenser valve position, concentration of product in byproduct stream.

6-3. Conversion, yield, product recovery, steam flow per unit product or feed.

6-4. Total operating cost.

6-5. Programmed optimization.

6-6. Boiler B, because its efficiency curve is still rising, while that of Boiler A is falling.

UNIT 7

7-1,2. Let F $= m_H + m_C$ (1)

$$T = \frac{m_H T_H + m_C T_C}{F} \quad (2)$$

combining,

$$F = \frac{m_H (T_H - T_C)}{T - T_C} \quad (3)$$

Differentiating (1),

$$\left.\frac{\delta F}{\delta m_H}\right|_{m_C} = 1$$

Differentiating (3),

$$\left.\frac{\delta F}{\delta m_H}\right|_{T} = \frac{T_H - T_C}{T - T_C}$$

Dividing,

$$\lambda_{FH} = \frac{T - T_C}{T_H - T_C} = \frac{50 - 10}{80 - 10} = 0.571$$

Arranging,

	m_H	m_C
F	0.571	0.429
T	0.429	0.571

7-3. From Eq. (3) above,

$$\frac{m_H}{F} = \frac{T - T_C}{T_H - T_C} = \lambda_{FH}$$

Therefore, λ_{FH} is the ratio of the hot-water valve capacity to the sum of the two capacities.

7-4. Opening each valve will raise its own flow and decrease the other flow, but by a smaller amount. $K_{11} = K_{22} > -K_{12} = -K_{21}$. Using Eq. (7-13), λ_{11} comes out to be greater than 1 but probably not greater than 2, since $-K_{21}/K_{11}$ is probably less than 0.7.

7-5. The 1−1, 2−2, 3−3 configuration is the best of a poor set of choices; the $c_2 - m_2$ loop will require decoupling from the others (see Unit 8).

7-6. The best pairing of single loops is: x − B, p − D, y − R, b − Q.

UNIT 8

8-1.. Water injection raises relative humidity and provides cooling—it should be used for humidity control. Outside air reduces temperature with little effect on **relative** humidity—it should be manipulated for temperature control.

8-2. Multiply m_1 by the output of controller 2 to develop m_2:

$$m_2 = m_1\, f(c_2)$$

Subtract m_2 from the output of controller 1 to develop m_1:

$$m_1 = f(c_1) - m_2K_2/K_1$$

This assumes $K_2 < K_1$; if $K_2 > K_1$, then control loop assignment must be reversed:

$$m_1 = m_2\, f(c_2)$$

$$m_2 = f(c_1) - m_1K_1/K_2$$

This keeps the decoupler gain below unity.

8-3. Only the decoupler for c_2 needs to be retained; it can be used in the form above, or

$$m_2 = F\, f(c_2),\ \text{if } K_2 < K_1$$

$$m_1 = F\, f(c_2),\ \text{if } K_2 > K_1$$

8-4. From Fig. 8-8:

$$m_1 = m_1' + m_2 J_{12}$$

$$m_2 = m_1 J_{21} + m_2'$$

Combining,

$$m_1 = \frac{m_1' + m_2' J_{12}}{1 - J_{12} J_{21}}$$

$$m_2 = \frac{m_1' J_{21} + m_2'}{1 - J_{12} J_{21}}$$

From Fig. 8-6,

$$c_1 = m_1 K_{11} g_{11} + m_2 K_{12} g_{12}$$

Combining the last three equations gives Eq. (8-29).

8-5. In the feedback configuration,

$$m_1 = f(c_3) - m_2$$

$$m_2 = f(\Delta c) + m_1$$

From an operational standpoint, the feedback decoupler makes less sense, in that valves m_1 and m_2 affect c_3 equally, yet would be manipulated by two different controllers.

8-6. Figure 1-4 shows a partial nonlinear decoupler; no dynamic compensation is needed because both flows enter at the same point. Figure 6-1 is a more elaborate version of 1-4, although the controller manipulating fuel flow is not shown.

UNIT 9

9-1. Yes, the feedback signal that develops integral action is limited to supply pressure.

9-2. A high-pressure limit would be achieved by connecting a pressure controller to the low selector; a low-level limit would be achieved by connecting a base-level controller to the same selector. Impending loss of temperature control is signaled by a deviation between TC output and feedback.

9-3. No—the suction valve is used to deliver the variable process load to the compressor. If the load were to fall to zero, due to a closed upstream valve, override of the suction valve could not keep the compressor out of surge.

9-4. Two analyzers would feed a high selector, whose output would go to an oxygen controller driving the oxygen valve. A high-temperature alarm would shut off oxygen flow through a separate valve, installed in series with the control valve.

9-5. The bed-temperature controller would send its output to the high selector setting air flow. This will raise air flow above fuel, thereby cooling the bed. However, because this is an inefficient operating mode, there should be an alarm whenever air flow stays above fuel flow for a significant time.

UNIT 10

10-1. Figures 2-7 and 2-8 both show variably structured systems. Pressure control is transferred from the turbine admission valve to the fuel flow controller in proceeding from startup to normal operating conditions. Separate pressure controllers are used.

10-2. Figure 3-7 shows three liquid levels capable of being controlled by either inflow or outflow; if inflow is constrained, outflow is manipulated; if outflow is constrained, inflow is manipulated. The use of separate controllers allows the capacity of the vessels to absorb temporary constraints without stopping production.

10-3. See Fig. B-10-1.

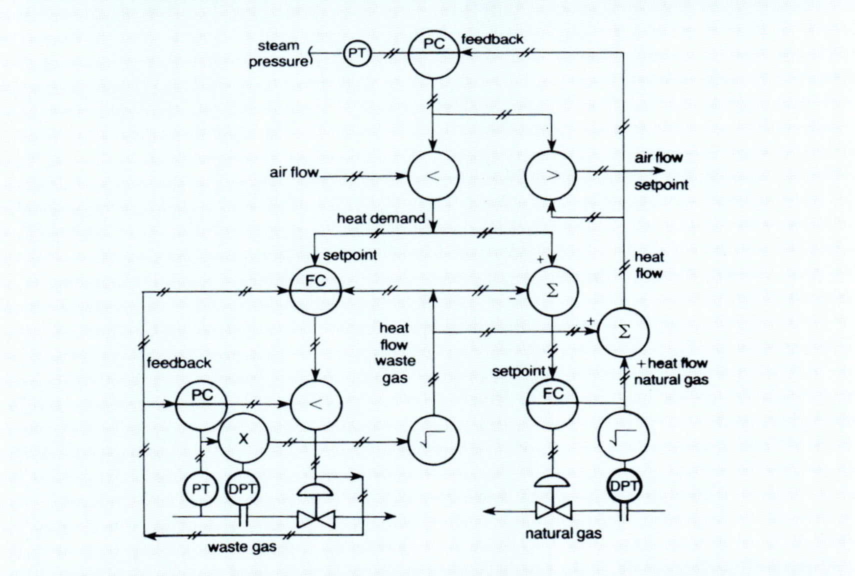

Fig. B-10-1. The heat that cannot be supplied by the waste gas flow controller is passed along to the natural gas.

10-4. The waste gas FC loop of Fig. B-10-1 is replaced by a wood feeder, whose setpoint may be reduced by an opacity controller through a low selector.

10-5. The compressors operate most efficiently when fully loaded, i.e., when their bypass valves are closed. Therefore they should be started in sequence under full load until the process demand is almost met. Then one additional unit should be started under part load, bypassed enough to meet the demand for refrigeration. Figure 10-6 applies.

UNIT 11

11-1. Differential pressure is measured across the bed and also across an orifice in the air duct. The ratio of these two differential pressures, as calculated by a divider, is proportional to the mass of solids in the bed.

11-2. Differential temperature between two tray locations is often used to infer composition; alternately, pressure and temperature measured at the same point are used to infer composition as vapor pressure or pressure-compensated boiling point. They both combine two controlled variables.

11-3. The rate of heat transfer is proportional to the log-mean temperature difference:

$$Q = UA \frac{(T-T_1) - (T-T_2)}{\ln \dfrac{T-T_1}{T-T_2}} = UA \frac{T_2 - T_1}{\ln \dfrac{T-T_1}{T-T_2}}$$

where T = reactor temperature
T_1 = coolant inlet temperature
T_2 = coolant outlet temperature
U = overall heat-transfer coefficient
A = heat-transfer area

The rate of heat transfer is also proportional to coolant flow and temperature rise: $Q = WC\,(T_2 - T_1)$

where W = coolant flow
C = coolant heat capacity

combining the above yields

$$\frac{T - T_1}{T - T_2} = e^{UA/WC}$$

Solving for T_2,

$$T_2 = T\,(1\text{-}e^{-UA/WC}) + T_1 e^{-UA/WC}$$

Since flow is constant,

$$T_2 = T^*(1-K) + T_1 K$$

where T^* is the desired reactor temperature, and K is a calibration parameter. The system operates just like the dryer control system of Fig. 11-6.

11-4. The feedforward system for the air compressor is identical to that of the boiler, shown in Fig. 11-7.

11-5. Figure 3-4 shows a system which inferentially controls the conversion of liquid feed by changing reaction pressure as a function of flow. It contains a positive-feedback loop requiring dynamic compensation.

Appendix C: Glossary of Control System Terminology

APPENDIX C

Glossary of Control System Terminology

NOTE: The definitions presented in this Appendix are taken (where applicable) from ISA Standard on Process Instrumentation Terminology (ISA-S51.1).

adaptive control—Control in which automatic means are used to change the type or influence (or both) of control parameters in such a way as to improve the performance of the control system.

antisurge control—Control by which the unstable operating mode of compressors known as "surge" is avoided.

available work—The capacity of a fluid or body to do work if applied to an ideal engine.

cascade control—Control in which the output of one controller is the setpoint for another controller.

conditional stability—The property of a controlled process by which it can function in either a stable or unstable mode, depending on conditions imposed.

constraint—The limit of normal operating range.

controlled variable—A process variable which is to be controlled at some desired value by means of manipulating another process variable.

dead band—The range through which an input can be varied without initiating observable change in output. (There are separate and distinct input-output relationships for increasing and decreasing signals.)

dead time—The interval of time between initiation of an input change or stimulus and the start of the resulting observable response.

dead zone—A predetermined range of input through which the output remains unchanged, irrespective of the direction of change of the input signal.

decoupling—The technique of reducing process interaction through coordination of control loops.

dew-point temperature—That temperature at which condensation of moisture from the vapor phase begins.

distillate—The distilled product from a fractionating column.

disturbance—An undesired change that takes place in a process that tends to affect adversely the value of a controlled variable.

dynamic compensation—A function which alters the dynamic response of an input signal to a process, to correct or counteract undesirable dynamic properties of the process.

endpoint control—The exact balancing of process inputs required to satisfy its stoichiometric demands.

feedforward control—Control in which information concerning one or more conditions that can disturb the controlled variable is converted, outside of any feedback loop, into corrective action to minimize deviations of the controlled variable.

flooding—That condition of a distillation column in which it begins to fill with liquid.

gain—The ratio of the change in output to the change in input which caused it.

humidity, absolute—The moisture content of air on a mass or volumetric basis.

humidity, relative—The moisture content of air relative to the maximum that the air can contain at the same pressure and temperature.

hydrocracker—A chemical reactor in which large hydrocarbon molecules are fractured in the presence of hydrogen.

incremental cost—The cost of the next increment of output from a process.

inverse response—The dynamic characteristic of a process by which its output responds to an input change by moving initially in one direction but finally in the other.

lag—The dynamic characteristic of a process giving exponential approach to equilibrium.

lead-lag compensator—A dynamic compensator combining lead action (the inverse of lag) with lag.

limit cycle—An oscillation of uniform amplitude.

limiter—A device which applies limits to a signal.

load—The demand for input to a process.

loop gain—The product of the gains of all the elements in a loop.

manipulated variable—A quantity or condition which is varied so as to change the value of the controlled variable.

multiple-output system—A system which manipulates a plurality of variables to achieve control of a single variable.

noise—an unwanted component of a signal or variable.

optimization—The act of controlling a process at its maximum possible level of performance, usually as expressed in economic terms.

partial pressure—The pressure exerted by a specified component in a mixture of gases.

pH—The negative base-ten logarithm of the hydrogen-ion activity in a solution.

phase shift—A change in phase angle between the sinusoidal input to an element and its resulting output.

positive feedback—A closed loop in which any change is reinforced until a limit is eventually reached.

primary loop—The outer loop in a cascade system.

proportional band—The change in input required to produce a full range change in output due to proportional control action.

reactor, endothermic—A reactor which absorbs heat from the surroundings.

reactor, exothermic—A reactor which generates heat.

reboiler—A heat exchanger on a distillation column which returns vapor to the column by boiling a portion of the liquid leaving it.

reflux—That portion of condensed distillate which is returned to the column.

regulate—The act of maintaining a controlled variable at or near its setpoint in the face of load disturbances.

relative gain—An open-loop gain determined with all other manipulated variables constant, divided by the same gain determined with all other controlled variables constant.

sample interval—The time interval between measurements or observations of a variable.

sampled-data control—Control action which is taken only momentarily following the measurement or observation of a variable.

secondary loop—The inner loop of a cascade system.

selector—A device which selects one of a plurality of signals.

self-regulation—The property of a process or machine which permits attainment of equilibrium, after a disturbance, without the intervention of a controller.

setpoint—An input variable which sets the desired value of a controlled variable.

sink—A reservoir into which material or energy is rejected.

source—A reservoir from which material or energy is drawn.

surge, in compressors—An unstable operating regime in which internal oscillations persist.

surge tank—A vessel used to absorb fluctuations in flow so that they are not passed on to other units.

time constant—The time required for the output of an exponential lag to travel 63.2% of its total excursion following a step change in input.

wet-bulb temperature—The adiabatic equilibrium temperature reached by a saturated wick in a moving stream of air.

windup—Saturation of the integral mode of a controller developing during times when control cannot be achieved, which causes the controlled variable to overshoot its setpoint when the obstacle to control is removed.

Wobbe Index –A function of heating value and specific gravity of a fuel gas which is needed to convert an orifice-meter differential pressure to units of heat flow.

Index

INDEX